Tim Peake
and Britain's Road To Space

Erik Seedhouse

TIM PEAKE
and BRITAIN'S
ROAD TO SPACE

 Springer

Published in association with
Praxis Publishing
Chichester, UK

Erik Seedhouse
Assistant Professor, Commercial Space Operations
Embry-Riddle Aeronautical University
Daytona Beach, Florida
USA

SPRINGER-PRAXIS BOOKS IN SPACE EXPLORATION

Springer Praxis Books
ISBN 978-3-319-57906-1   ISBN 978-3-319-57907-8   (eBook)
DOI 10.1007/978-3-319-57907-8

Library of Congress Control Number: 2017950171

Cover design: Jim Wilkie
Project Editor: Michael D. Shayler

Printed on acid-free paper

This Springer imprint is published by Springer Nature
The registered company is Springer International Publishing AG
The registered company address is: Gewerbestrasse 11, 6330 Cham, Switzerland

Contents

Acknowledgements

In writing this book, the author has been fortunate to have had five reviewers who made such positive comments concerning the content of this publication. He is also grateful to Maury Solomon at Springer and to Clive Horwood and his team at Praxis for guiding this book through the publication process. The author also gratefully acknowledges all those who gave permission to use many of the images in this book and also to Project Manager, Sasi Reka, for guiding this manuscript through the final stages of publication. The author also expresses his deep appreciation to Michael Shayler, whose attention to detail and patience greatly facilitated the publication of this book and to Jim Wilkie for creating yet another striking cover. Thanks Jim!

To
Alice

About the Author

Dr. Erik Seedhouse [from the author's collection]

Erik Seedhouse is a fully-trained commercial suborbital astronaut. After completing his first degree, he joined the 2nd Battalion the Parachute Regiment. During his time in the 'Paras', Erik spent six months in Belize, where he was trained in the art of jungle warfare. Later, he spent several months learning the intricacies of desert warfare in Cyprus. He made more than 30 jumps from a Hercules C130 aircraft, performed more than 200 helicopter abseils and fired more light anti-tank weapons than he cares to remember!

Upon returning to academia, the author embarked upon a Master's degree which he supported by winning prize money in 100 km running races. After placing third in the World 100 km Championships in 1992, Erik turned to ultra-distance triathlon, winning the World Endurance Triathlon Championships in 1995 and 1996. For good measure, he won the World Double Ironman Championships in 1995 and the infamous Decatriathlon, an event requiring competitors to swim 38 km, cycle 1800 km, and run 422 km. Non-stop!

In 1996, Erik pursued his PhD at the German Space Agency's Institute for Space Medicine. While studying, he found time to win Ultraman Hawai'i and the European Ultraman Championships, as well as completing Race Across America. Due to his success as the world's leading ultra-distance triathlete, Erik was featured in dozens of magazine and television interviews. In 1997, GQ magazine named him the 'Fittest Man in the World'.

In 1999, Erik took a research job at Simon Fraser University. In 2005, the author worked as an astronaut training consultant for Bigelow Aerospace. Between 2008 and 2013, he served as Director of Canada's manned centrifuge and hypobaric operations. In 2009, he was one of the final 30 candidates in the Canadian Space Agency's Astronaut Recruitment Campaign. Erik has a dream job as an assistant professor at Embry-Riddle Aeronautical University in Daytona Beach, Florida. In his spare time, he works as an astronaut instructor for Project PoSSUM, an occasional film consultant to Hollywood, a professional speaker, a triathlon coach and an author. 'Tim Peake' is his 27th book. When not enjoying the sun and rocket launches on Florida's Space Coast with his fiancée, Alice, he divides his time between his second home in Sandefjord and Waikoloa.

Acronyms and Abbreviations

ACE	Advanced Colloids Experiment
AES	Advanced Exploration Systems
ATA	Ammonia Task Assembly
ATV	Automated Transfer Vehicle
BEAM	Bigelow Expandable Activity Module
BEO	Beyond Earth Orbit
BIS	British Interplanetary Society
BNSC	British National Space Centre
BRT	Body Restraint Tether
CAVES	Cooperative Adventure for Valuing and Exercising human behavior and performance skills
CETA	Crew and Equipment Translation Aid
CIDS	Circuit Interrupt Devices
COLBERT	Combined Operational Load Bearing External Resistance Treadmill
COSPAR	Committee on Space Research
CPS	Consolidated Planning System
CSA	Canadian Space Agency
CSF	Cerebrospinal Fluid
DCM	Display Control Module
DCS	Decompression Sickness
DCSU	Direct Current Switching Unit
DTI	Diffusion Tensor Imaging
EDR	European Drawer Unit
ELDO	European Launch Development Organization
ELIPS	European Life and Physical Sciences
EMU	Extravehicular Mobility Unit
EPFTP	EVA Pre-Familiarization Training Program
EPM	European Physiology Module

ESRC	Economic and Social Research Council
ETC	European Transport Carrier
EVA	Extravehicular Activity
FPS	Fan Pump Separator
FSL	Fluid Science Laboratory
GCR	Galactic Cosmic Radiation
GNC	Guidance Navigation and Control
HPA	Hypothalamic Pituitary Adrenal
HST	Hubble Space Telescope
HUT	Hard Upper Torso
ICC-VLD	Integrated Cargo Carrier-Vertical Light Deployable
ICP	Intracranial Pressure
IDA	International Docking Adapter
IGA	Intergovernmental Agreement
ISLE	In-Suit Light Exercise
ISS	International Space Station
IVA	Intravehicular Activity
JAXA	Japanese Space Agency
JEM	Japanese Experimental Module
KSC	Kennedy Space Center
LCVG	Liquid Cooling Ventilation Garment
LEO	Low Earth Orbit
LET	Launch Escape Tower
LMM	Light Microscopy Module
MAG	Maximum Absorbency Garment
MMOD	Micrometeoroid Orbital Debris
MRI	Magnetic Resonance Imaging
MRM	Mini Research Module
NBL	Neutral Buoyancy Laboratory
NEEMO	NASA Extreme Environments Mission Operations
NOAA	National Oceanic and Atmospheric Administration
NPV	Non-Propulsive Vent
OSTPV	Onboard Short-Term Plan Viewer
PARE	Physiological and Anatomical Rodent Experiment
PLLS	Primary Life Support System
PMA	Pressurized Mating Adapter
RCS	Reaction Control System
SAM	Surface-to-Air Missile
SM	Service Module
SPDM	Special Purpose Dexterous Manipulator
SRB	Solid Rocket Booster
SSAS	Segment-to-Segment Attachment System
SSDS	Surface Supplied Diving System
SSU	Sequential Shunt Unit
STEM	Science, Technology, Engineering and Mathematics

T-RAD	Tile Repair Ablator Dispenser
TCU	Thermal Control Unit
TEPC	Tissue Equivalent Proportional Counter
TPS	Thermal Protection System
TVIS	Treadmill with Vibration and Isolation Stabilization
UKSA	United Kingdom Space Agency
ULA	United Launch Alliance
VIIP	Visual Impairment Intracranial Pressure
XBASE	Expandable Bigelow Advanced Station Enhancement

List of Tables

Preface

"Tim will continue with his training at the European Space Agency, but if we don't fund any more then he won't get a second flight. We don't lose all the science that we have done, we don't lose the enthusiasm of the young people for science – but where are we in the UK? Just another backward nation that is not participating on the international stage, in the future of the human race? We are a travelling nation; we have explored for centuries and I think it's almost in our blood. It is something that Britain wants to do, Britain needs to do and we have got to continue that funding."

Helen Sharman, on the eve of her 25th anniversary
as Britain's first astronaut.

For centuries, the British developed a reputation as a nation of explorers. From Francis Drake's circumnavigation of the globe to the British-sponsored ascent of Mount Everest, British explorers crossed oceans and continents and ventured where few, if any, had gone before. But until very recently, that legacy of exploration had not extended to space. Most other major space-faring nations have shown at least some level of interest in human spaceflight, either by developing their own capability of sending humans into orbit (the U.S., Russia and China), or by piggybacking on those programs (Canada, Japan, and several major European nations). Sadly, the British government (Margaret Thatcher in particular) has had a long-standing opposition to such efforts, declining either to contribute to ESA's human spaceflight efforts or to fund its own. For decades, successive British governments chose to stay out of ESA's human spaceflight program, looking on as more than half a dozen other European countries sent astronauts into Earth orbit. But in 2008, there were signs of optimism for supporters of a UK government-sponsored human space effort when ESA selected a new class of six astronauts. For the first time, the new group included a British representative: Timothy Peake. Then, *finally*, the incumbent coalition government committed £27 million to ESA in 2012 and a further £49 million in 2014. These contributions paved the way for Tim Peake's mission to the International Space Station (ISS). The rest, as they say, is history, and is chronicled in the book you are reading.

> "It's absolutely clear why he got selected against all the odds. They couldn't let him go. He stands out. Most astronauts now are similar, but he is up there in the exceptional class. I think that there was a cowboy element to the original astronauts. Not so with Tim. Tim is ever cool, calm and collected whilst always seeming more charming than macho. When you talk to him, that calmness comes through. It is just about possible to believe that he really enjoyed himself facing the challenge of exercises like escaping from a helicopter cockpit whilst suspended upside down in water. I'd say he is made not so much of the right stuff but rather 'even better stuff'."
>
> *David Southwood, senior researcher at Imperial College,*
> *and member of the UK Space Agency steering board.*
> (The Guardian, *Ian Sample, December 11, 2015*)

The aim of this book was to put you, the reader, in the flight suit of Britain's first male astronaut. To that end, this book takes you on the journey followed by a British Army officer from Chichester, who spent over 185 days living and working on the ISS, including four hours and 43 minutes of space walks. This book is not organized like the typical biography, which usually follows the chronology of the subject from childhood to the present day. Instead of following a linear timeline, this book uses Tim Peake's experiences to illustrate larger points and themes, such as the stresses of an intensely scrutinized, highly visible job, the challenges of extended family separation, and the ever-present possibility of having to make the ultimate sacrifice: All this against the back-drop of a story-hungry press, starved of any news of British astronauts for the best part of 25 years.

Credit: ESA

"I think it is really important to reach out to our younger generation and to try to encourage them to take up science, technology, engineering and mathematics as subjects. We have a skills shortage at the moment, and we desperately need more graduates with those backgrounds."

Tim Peake, speaking with The Mail Online,
in an interview with Victoria Woolaston, November 6, 2015.

Also discussed are the learning curves that have to be met during astronaut and mission training, and the complexity of the technologies required to launch an astronaut into space and keep them alive for months on end. The narrative in the book is written in a way that allows the story and the people to propel the book. The rationale for taking this approach, as opposed to writing subject by subject, is that technology and training, unlike space, does not exist in a vacuum. Complex technical systems, such as the ISS, interact with the variables of human personality, the cultural backgrounds of the astronauts and cosmonauts, and, indeed, the 'culture' which permeates organizations like the British government and ESA.

Credit: ESA

"Obviously, this is the first time a UK Astronaut has flown on board the ISS as part of the European Space Agency, so that's the big thing here. From a government perspective, the UK is becoming involved in human space flight and it is something very important. I hope we go on to continue this involvement. So yes, it is very important the UK is part of this, as there is so much benefit to be had from ISS research in terms of what we are doing on the ISS for people back on Earth. And also on the ISS for future exploration – looking ahead to those lunar and Mars missions, and deeper into the solar system as we go on. I don't want the UK to miss out on that. ESA has been doing a fantastic job, and will continue to do a fantastic job in human spaceflight. I think it is definitely time for the UK to be part of that, and continue to be part of it. It's only going to get bigger and better."

Tim Peake, in an interview with RocketStem's Sam Mundell,
February 16, 2015

Credit: ESA

In addition to delving into the life and career of Tim Peake, this book weaves into the narrative the tortured and intransigent political history of manned spaceflight in the UK. Tim Peake's flight was an opportunity that had existed for decades, but was one that had been perpetually almost micro-managed and mismanaged out of existence by successive generations of myopically-minded politicians and bureaucrats. Along the way, the book aims to correct the myriad misunderstandings and warped impressions the British public have about the program: basically, correcting decades of sound-bites made by the sometimes spectacularly misinformed tabloid press. But ultimately, this book is the story of Tim Peake and the Principia mission – and the down-to-the-last-bolt descriptions of life aboard the ISS – by way of the hurdles placed in the path by the British government and the rigors of training at Russia's Star City.

"The younger generation that I'm talking to as I tour around the UK really will see humans land on Mars for the first time, which is incredibly exciting. We're now looking to set up a habitational module on the Moon as a stepping stone for Mars. I've just been amazed at how well we can live and work in space and still come back to Earth in great shape.

Credit: ESA

"When I've been speaking to my friends on board the ISS now, I do miss it. I miss the view of the Earth and I miss weightlessness – so yes, if the chance came up to start a habitational module on the Moon, I'd be first in the queue."

Tim Peake, talking on Spirit FM, October 18, 2016,
on the subject of a second mission and becoming
the first Brit to set foot on the Moon.

1

26 million horsepower

Figure 1.0: Credit: ESA

DECEMBER 15, 2015, BAIKONUR COSMODROME, KAZAKHSTAN

The area around the Soyuz rocket is cleared for more than a mile. Clustered in the observation area are family and friends of the crew. Watching a rocket launch from a mile away
reduces neck strain and is easier on the hearing when the roar of 26 million horsepower is

© Springer International Publishing AG 2017
E. Seedhouse, *Tim Peake and Britain's Road To Space*, Springer Praxis Books,
DOI 10.1007/978-3-319-57907-8_1

unleashed. But the main – and more sobering – reason for this separation is crowd safety: if the Soyuz *does* explode, the onlookers should be out of harm's way. But that scenario is probably not on the mind of Britain's first astronaut in almost a quarter of a century. This is not because he is oblivious to danger, but simply that he is more accustomed to it than most. After all, that's what test pilots do: they approach risk as something that must be understood, and they do that through a process of training and preparation which is not that dissimilar to astronaut training. So, for all the inherent dangers in the risky business of manned spaceflight, Major Peake's (Figure 1.1) mindset is not focused on a conflagration but rather on the myriad potential problems that may prevent the rocket from launching. Fear? Not likely. This is someone who is as calm and level-headed as they come. An experienced pilot who has flown reconnaissance missions over Northern Ireland and Afghanistan and over the mountains of Bosnia. At night. In winter. Over terrain littered with mines.

> "Tim was one of those guys who stood out in so many ways, even when he was a captain. I wasn't a bit surprised that he was selected. He's modest, he's highly talented, he is a hugely professional aviator and he's a great test pilot. He's the perfect guy to be representing the UK and to be representing the armed forces."
>
> *Richard Folkes, former director of British Army Aviation,*
> *who has known Tim Peake for more than a decade.*

Figure 1.1: Tim Peake. Credit: ESA

While fear may not have been uppermost in Major Peake's mind as he prepared for launch, the prospect of 270 tonnes of burning kerosene propelling a 26 million horsepower rocket with him sitting on top of it was probably of concern to his wife Rebecca and their two sons. That anxiety was probably also felt by Major Peake's parents, Angela and Nigel, who waited in the VIP observation area watching the launch timeline scroll down to zero. They were no doubt hoping that all the prelaunch traditions Tim and his colleagues had gone through would pay off in the form of a flawless launch. You see, the Russians are a superstitious bunch when it comes to spaceflight. Not that you can blame them. After all, if you're sitting on top of nearly 300 tonnes of rocket fuel, you probably don't want to push your luck by tripping over any proverbial black cats. Having said that, the Russians tend to take the business of tradition more seriously than most. Consider the signing ritual, for instance. If you're taking a trip to orbit from Russia then you need to make sure you bring a pen, because there are lots of things that must be signed. To begin with, the crew must sign off their spacecraft to approve their vehicle. This may sound strange because none of the crew has had any part in its construction, so it is not clear what the repercussions would be if the crew didn't approve their vehicle. After signing off the Soyuz, the crew must then sign the wall in the Baikonur museum *and* the door of their bedroom (Figure 1.2) before their last night on *terra firma* (when they return, the crew also signs the casing of their capsule).

Figure 1.2: Traditional door signing before launch. It is customary for each crewmember to sign their name on the door of their bedroom after their last night on Earth. Credit: ESA-S. Corvaya.

Another ritual is making sure that you are far removed from the vehicle when it is being rolled out to the pad, because the Russians consider it bad luck for the crew to see their vehicle being rolled out. And on the subject of roll out, as the Soyuz is transported from its hangar by rail, onlookers place coins on the track to be flattened by the train as it inches its way to the pad. This is to bring good luck to the mission and as far as we know, this practice has never led to any derailments, which is probably one reason why this ritual has continued.

As launch day approaches, the rituals come thick and fast. Before the launch, the crew must watch *The White Sun of the Desert*, a 1970 movie that has been watched by *every* crew that has flown from Baikonur. On their way to the pad, the crew must also make a pit stop to relieve themselves, a tradition that goes right back to Yuri Gagarin on his way to the launch of the very first human space flight (he urinated on the rear right tire of the transport bus, for those who need to know the details of this sort of stuff). While this made sense back in the 1960s, given the limitations of spacesuit design in those days, there is no rhyme or reason to the custom today, since the suits are now fitted with diapers. Still, you don't mess with tradition, so today's spacefarers are still expected to unzip and aim their appendage at the right rear wheel. Afterwards, the suit techs do their thing and have to go through the whole rigmarole of zipping the astronauts back in their suits. And if you happen to be a female astronaut? Well, then you bring a sample along with you in a vial and splash it on the wheel.

Signed your name? Check. Watched the film? Check. Urinated on the bus? Check. Great. After he had performed all the necessary pre-ingress rituals, it was time for Peake to clamber on board the 50-meter rocket. Once ensconced in their Soyuz, the crew methodically checked off their pre-flight checklist before attending to yet more pre-flight customs. One of these was to listen to Russian love songs, another tradition started by Gagarin (who else?) who, having checked off all his 'to do' items, requested that Mission Control play some music. The same happened with Peake's flight, although the crew (Figure 1.3) were permitted a selection of their own songs. Peake's selection? *Don't Stop Me Now* (by Queen), *Beautiful Day* (U2), and *A Sky Full of Stars* (Coldplay). Another rite of passage before the crew can get off the ground is to hang a cuddly toy from the instrument panel. This cuddly toy – which the astronauts usually let their kids choose – serves a dual purpose, both as a mascot for the mission, and as a way of indicating to the crew that they have achieved weightlessness. By now you may be wondering if all these traditions aren't a little over the top, but who's to say there isn't something to them. After all, let's not forget that no other country has such an incredible safety record when it comes to the business of sending humans into space: Since 1961 the Russians have lost just 4 cosmonauts during a mission (Soyuz 1 pilot Vladimir Komarov in 1967, and the 3-man Soyuz 11 crew in 1971). The U.S. program, meanwhile, has lost two Shuttle crews (7 per crew in 1986 and 2003) and a SpaceShipTwo pilot. But back to the launch.

GO for flight! Thanks for all the good luck messages – phenomenal support! #Principia https://t.co/8jbxejHEEe

Tim Peake's last tweet before launch

Figure 1.3: Expedition 46 crew (front to back) Tim Peake (ESA), Tim Kopra (NASA), Yuri Malenchenko (Roscosmos). Credit ESA

A volcano of flame erupts from the engines (Figure 1.4). A second later, an ear-splitting roar is heard by the observers in the VIP area. The clock is running on the Principia Expedition. Sheets of ice fall away. Umbilicals are ripped from their tethers. The gantries jerk back and the giant hold-down arms release the Soyuz as it begins to crawl skyward on a plume of flame. Moments later, the fiery rocket is muscling its way through the upper reaches of the atmosphere. Inside the capsule, Peake and his crewmates, their breathing fast and shallow, perform anti-G straining maneuvers as they are pummeled by Gs. The instrument panels scroll the numbers, the velocity accelerating at an incredible rate. In the VIP area, Peake's family and friends look skyward at the rocket, trailing thick smoke, as it arcs eastwards towards the horizon. The sound is still deafening, even with the rocket dozens of kilometers downrange. Inside the Soyuz, the crew continue to be buffeted as the engines adjust the trajectory to give chase to the International Space Station (ISS) which, at their current speed of more than 28,000 kilometers per hour, is just six hours away. Soon, very soon, the only evidence to be seen of a rocket launch is a smoky white contrail curling softly away against the blue sky.

Figure 1.4: Launch of the first Brit in space for 25 years. In among the smoke and flames is a 50-meter-high launcher consisting of three sections. The launcher delivers 26 million horsepower to reach an orbital speed of 28,800 km/h. Within ten seconds of rising from the pad, the three men inside will have travelled over 1,640 km and they will reach the International Space Station in six hours. Helping them to travel at these dizzying speeds are 300 tonnes of propellant; a mixture of kerosene and liquid oxygen generates the propulsion. Credit ESA

OPERATION *BACKFIRE* - THE BRITISH SPACE PROGRAM V 1.0

It was a cold autumn day in 1945 when German soldiers of the *Altenwalde Versuchskommando* (AVKO) readied their V-2 rocket. It was a procedure they had performed countless times before, because the V-2 was the rocket that had been used by the Germans to terrify the people of London in the latter part of Second World War. Preparations complete, the officers and soldiers looked on as the V-2 roared skyward, rapidly disappearing into the distance. The AVKO men managed a smile and a subdued cheer and were soon joined ... by a group of *British* officers who accompanied them in the celebrations. Thus began the genesis of what would become known as the first British Space Program.

Thanks in part to Hollywood, when most people look back to the space race between the U.S. and the USSR they think of it as a two-horse race that became known as Operation *Paperclip*. The reality is somewhat different, because Great Britain was just as aware of the significance of this rocketry business as the two superpowers. Part of that understanding came from the unfortunate experience of being one of the very few countries in the world whose cities have been attacked by ballistic missiles. Unfortunately, in the chaos that ensued in the race to grab as much of Germany's rocket knowhow as possible, Great Britain lost out. The U.S. walked away with Wernher von Braun, together with priceless blueprints and documents (many of which had been liberated from the British sector) that the great German had saved from destruction as the Nazi system imploded.

A BRITISH MAN IN SPACE. BY 1950!

How this happened, and why the British let it happen, is not well understood, but what *is* known is that the British were just as interested in the practicalities of German rocketry as the Americans and the Russians, which is how Operation *Backfire* was born. Following the end of the Second World War, the British searched for soldiers from the V-2 units and assembled them into what became known as the AVKO. Almost overnight, the same soldiers who had once launched lethal ballistic missile attacks on London were now working shoulder to shoulder with those who had been on the receiving end of all that mayhem and destruction. As the German and British soldiers worked together, the British gained valuable knowledge of high-test peroxide rocketry, knowledge that was needed for Britain's nuclear deterrent. Remember, this was the time when the 'Iron Curtain' of the Soviet Bloc was descending at the beginning of the Cold War and rocketry was seen as one deterrent against any escalation of the threat on both sides, even if the grander efforts of Korolev and Von Braun (who had long since been spirited away to the U.S.) were out of the financial range of the British post-war budget.

At its core, Operation *Backfire* involved launching V-2 rockets from the Netherlands to the edge of space, to test the accuracy of the German weapon. These launches were successful, since the rockets landed within five kilometers of their targets, providing evidence that Von Braun had solved the basic challenges of rocketry. At the same time as these rocket launches were taking place, engineers at the British Interplanetary Society (BIS) reckoned that Von Braun's technology could be re-purposed to build a spacecraft: a man-rated

Table 1.1 British V-2 Launches

Date	Time	Maximum height	Length of flight
October 2, 1945	14:41	69.4 kilometers	249.4 kilometers
October 4, 1945	14:16	17.4 kilometers	24 kilometers
October 15, 1945	15:06	64 kilometers	233 kilometers

V-2, in other words. So the engineers, led by R.A. Smith, went to work modifying a V-2, by strengthening the hull, increasing the fuel payload and replacing the warhead with a capsule capable of carrying a human. While the design, which was known as Megaroc, would not have been capable of launching a human into orbit, the rocket would have been powerful enough to place a human in space at the apex of a high altitude parabolic trajectory. Impressive stuff, in 1946!

In practical terms, the rocket would have been launched at an angle of two degrees. Once in space, a segmented nose would have separated to reveal a capsule. Inside the capsule would have been the world's first (British!) astronaut, who would probably have used his few minutes of weightlessness (this was a parabolic/suborbital flight, remember) to conduct observations of the Soviet Union. During re-entry, the capsule's heatshield would have protected the historic astronaut and parachutes would have deployed to ensure a soft landing.

"The design was totally practical. All the technology existed and it could have been achieved within three to five years."

David Baker, space historian and editor of Spaceflight *magazine,*
who has studied the British V-2 designs

Too ambitious? Not according to engineers who studied the design. David Baker, quoted earlier, was trained on V-2 technology and spent much of his career as a NASA Space Shuttle engineer, so he happens to know a little about what makes rockets do what they do. According to Baker, the British V-2 effort (see Table 1.1) was ten years ahead of its time; in other words, Britain could have launched the world's first astronaut in 1951. Instead, they had to wait another 40 years. R.A. Smith submitted his design to get Britain into the manned spaceflight business to the British government in 1946, but it was rejected. The reason? Britain was practically bankrupt after having spent all its money winning the war. Technically, the British could have beaten Gagarin, but financially, Smith's plan couldn't have come at a worse time.

On the other side of the pond meanwhile, the situation was very, very different. The U.S. gave Von Braun unlimited funds to develop the V-2 into what eventually became the Mercury-Redstone, the rocket that carried Alan Shepard into space in 1961 (and which was eerily similar to Smith's design). So, in an alternate reality in which the British government had been flush with money, the world's first astronaut might well have been British. Instead, the British revamped V-2 project was consigned to the 'what might have been' dustbin of history and for decades the spirit of Operation *Backfire* was forgotten, as successive governments squandered opportunity after opportunity to show what the British could do in space if given half a chance. But that isn't to say they had given up on rocketry, because after Operation *Backfire* came Blue Steel.

BLUE STEEL

While Blue Steel was never designed with launching humans in mind, the project added another level of rocketry knowhow to Britain's experience in the business of building things that flew high and fast. The catalyst for Blue Steel was a Ministry of Supply memo, dated November 5, 1954, which warned that by 1960, Soviet air defenses would make it dangerous for Britain's V bombers to attack with nuclear weapons. The solution was a rocket-propelled supersonic missile, with a range of at least 80 kilometers and the capability of carrying a nuclear weapon. With these parameters, the bombers would be able to remain out of the range of Soviet ground-based defenses, but would still be capable of firing a weapon that could hit Soviet targets very quickly, thanks to its Mach 3 capability.

Less than 10 years after the memo was written, Blue Steel entered service. By this time, however, the improved surface-to-air missiles (SAMs) that the Soviets had developed were capable of a much greater range, which all but cancelled out any advantages of the Blue Steel design. It was suggested that an upgraded Blue Steel II should be built to overcome this, but that plan was dismissed in favor of the U.S. (AGM-48) Skybolt system. Unfortunately, the Skybolt system was cancelled, leaving the British with no option but to operationalize Blue Steel. So, despite its shortcomings, Blue Steel entered service in 1963.

A winged vehicle with clipped delta wings, the missile (which took seven hours to prepare for launch) was powered by an Armstrong Siddeley Stentor Mark 101 rocket engine that burned hydrogen peroxide and kerosene and produced 24,000 lb. of thrust. Carried by Avro Vulcan and Handley Page Victor bombers, the high-altitude weapon was capable of flying at Mach 3 but proved unreliable, and the program was retired at the end of 1970. While Blue Steel was a troubled program, the British did gain more – albeit painful – experience in dealing with rockets and the awkward hydrogen peroxide fuel. That said, the program didn't really deliver a step forward in Britain's space program, but that's not to say the British had given up on the idea of a true British space rocket. Enter Blue Streak.

ONCE UPON A TIME IN THE 60S

Blue Streak was another of the Cold War's experimental missiles, one that ultimately became the first stage of the European space launch system – more about that later. The impetus to begin developing the vehicle emerged within a year of the 1954 Blue Steel memo. Designated as a Medium Range Ballistic Missile, with a range of 3700 kilometers, Blue Streak was designed to carry a nuclear warhead that could be dropped on the Soviet Union. Today, the weapon would be known as a Weapon of Mass Destruction (WMD). While the term 'Blue Streak' may sound like an appropriate codename for a project with spaceflight aspirations, the name does not actually mean anything. It was just a name taken randomly from a post-World War 2 book of 'Rainbow Codes' (other codenames included 'Brown Bunny' and 'Orange Poodle'!). Since strategic missile projects were designated as 'Blue' projects, the codenames Blue Steel and Blue Streak fit the bill nicely.

The Blue Streak project had a long lead time because much of the technology was cutting-edge and had yet to be developed, and almost inevitably, as with so many new technology projects, its costs spiraled out of control. Lacking the bottomless military budget of the Americans, the British had no other option but to cancel the project in 1960. But this cancellation only applied to the use of Blue Streak as a weapon (the Polaris and Trident missiles ultimately performed the deterrent role that Blue Streak was slated for), not as an element in the space program, which was being pursued by the British very seriously at the time. It was decided to repurpose Blue Streak as the first stage of a planned British three-stage rocket known as Black Prince (rockets were given a 'Black' designation in the Rainbow Codes; another rocket program was Black Knight, which was used to test the Blue Streak system).

Ultimately, Black Prince was not built, and instead the British decided to work together with other European countries by forming the European Launcher Development Organization (ELDO). This was a partnership between Britain, France, Germany, Belgium, Holland and Italy, together with non-European associate member, Australia, which had a suitably deserted test site at Woomera. The ELDO's first rocket, Europa 1, used the Blue Streak as the first stage (along with a second stage built by the French and a third stage by the Germans). The first stage was launched four times from Woomera, but when it came to combining the first, second and third stages, the outcome fell far short of expectations, with all eight combined test vehicles being destroyed. The British government, understandably fed up with all the problems, grew impatient and cancelled Blue Streak. With an American-built variant taking the place of Blue Streak, the British government worried about a backlash from the public about wasting funds and decided to announce another program that would put satellites in space, and so Black Arrow (Figure 1.5) was born. But building rockets was now out of favor with the British Treasury, which ruthlessly reduced the budget for the new program, with the result that only five Black Arrows were ever built and only one satellite launch attempt was made.

After cannibalizing the technology for Blue Streak and Black Knight, engineers produced a rocket that could place a satellite into Polar orbit – on paper at least. By 1969, the first of the Black Arrows was ready to launch from Woomera, but the vehicle barely made it off the launch pad before it was detonated for safety reasons. The second launch was a success, so the engineers gambled on launching a satellite on the third vehicle (they had little choice because there was no money left). All went well until a leak in the pressurization system caused the propulsion to cut out and the satellite failed to make it into orbit. More bad news followed when the project was cancelled, and with it Britain's quest for space. Well, not quite. There was one last hurrah, because when the project's cancellation had been announced, the fourth Black Arrow was already on its way to Woomera. Since it would have cost more to bring it back than it would to proceed with the launch, the Treasury reluctantly agreed to allow the Black Arrow team to roll the dice one more time. On October 28, 1971, the fourth Black Arrow sat on the launch pad with a satellite – Prospero – perched on top of its third stage. This time, all went well, and Britain joined the very select group of countries that had independently launched a satellite. Sadly, on the same day, the UK also became the first and only country to relinquish its membership of this very eclectic group. And what of Prospero? The satellite is still up there, a relic of a bygone era.

Figure 1.5:   A Black Arrow in Woomera Rocket Park. Britain's only indigenous launch vehicle, the project was cancelled in July 1971.
Technical specifications: *First Launch*: June 27, 1969. *Last Launch*: October 28, 1971. *Payload*: 73 kg. *Thrust*: 222.40 kN. *Gross mass*: 18,130 kg. *Height*: 13.00 m. *Diameter*: 1.98 m. *Apogee*: 200 km.
The rocket, which used a unique propellant combination, was a three-stage vehicle. The first stage sported an 8-chamber layout, with four pairs of Gamma engines gimballed to provide thrust in all axes. The two Gamma chambers of the second stage were also gimballed, to provide thrust in three axes. The third stage was a Waxwing solid rocket motor. Credit: Magnus Manske

ARIEL-1

Dovetailing the development of Blue Streak was a slightly more successful program that was kick-started by a 1959 meeting of the Committee on Space Research (COSPAR). At that meeting, the United States offered to launch scientific satellites built by foreign countries, an offer the British responded to via the British National Committee on Space Research, which made its proposal to NASA. And so, Ariel-1 was born. While NASA agreed to build and launch the vehicle, the British developed the experiments (see Table 1.2), which were designed to investigate the relationship between solar radiation and the ionosphere.

Table 1.2. Ariel-1 Experiments [1]

Ionosphere Experiments
- Langmuir probe for measurement of electron temperature and density
University College London
Determine the value of the electron density and temperature near the satellite.
- Spherical probe for measurement of ion mass composition and temperature
University College London
Similar to the electron temperature experiment but using a different method of temperature
  measurement.
- Plasma dielectric constant measurement of ionospheric electron density
University of Birmingham
Provide electron density measurements similar to the UCL experiment but using a different and
  complementary method.

Solar radiation experiments
- Measurement of solar Lyman-Alpha emission
University College London
Monitor the intensity of radiation in the ultra-violet (Lyman-Alpha) range of the solar spectrum
- Measurement of the X-Ray emission from the Sun in the 3 to 12A band
University College London
Provide an indication of solar conditions by monitoring the intensity of radiation in the ultra-
  violet (Lyman-Alpha) range of the solar spectrum.

Cosmic Ray experiment
- Cosmic Ray Analyzer
Imperial College London
Using a Cerenkov detector provide accurate measurements of the cosmic ray energy spectrum
  and the impacts of interplanetary magnetic field modulation on this spectrum.

Ariel-1 (also known as UK-1 and S-55) became the first British satellite – predating
Prospero by almost ten years – when it was launched on top of a Thor-Delta rocket in April
1962 (Figure 1.6). The mission went well for a while, but in September that year, a high-
altitude nuclear test caused damage to the satellite's solar panels and its performance was
significantly affected. The irony of this event was that the nuclear test in question – Starfish –
was American! For history buffs, the event marked a space first of sorts because it represented
the first time a superpower had destroyed a satellite flown by its own space agency. The quick
end to the world's first international satellite was later spelled out in declassified documents
originally published by the British Office of the Minister for Science, as follows:

"Everything went well from launch for several weeks and we were getting very
interesting data. We showed how the X-ray flux from the Sun changed dramatically
when there was a solar flare. That was cutting-edge science at the time and was a
kick-off to solar weather, which is now a mini industry. But then everything changed.
Suddenly, without any warning, our X-ray count rate went off the scale. Everything
went wild. We knew there was something strange going on but our initial suspicion
was that our detector was suffering from electrical breakdown."

Ken Pounds, Emeritus Professor (space science), Leicester University,
recalling the events of 1962 and Ariel's untimely demise
(Prof. Pounds was in his 20s at the time).

Figure 1.6: Ariel-1 satellite. The 62-kilogram satellite was designed to enhance understanding of the ionosphere. It carried a tape recorder and instrumentation to detect cosmic-rays and solar emissions. Credit: Nick Stevens

"It seems incredible now that they carried out atmospheric nuclear tests, but the huge cloud of radiation and particles created was spread all around the Earth and affected quite a few spacecraft, including Ariel-1. Basically, it degraded Ariel-1's ability to generate its electric power from its solar generators, so it began to behave badly and eventually died. The other instruments were also experiencing unusual readings. We discovered after just a few days that there something strange going on because there were these mysterious communications from the other side of the Atlantic. There was a call that came through to the UK site from NASA asking, rather enigmatically, 'Have you seen anything strange lately?' It wasn't an immediate discovery that the Starfish tests had caused the problems with Ariel-1."

David Parker, of the UK Space Agency

The contents of these papers state quite clearly that NASA had no idea what the USAF was up to. In fact, NASA was so far out of the loop that the agency launched a Telstar communication satellite the day after the Starfish event! The cost was not insignificant, since NASA had spent $2 million compared to the UK's £200,000. The British reaction to the Ariel-1 affair is summed up (in a Shakespearean tone that echoed the British penchant for giving their satellites Shakespearean names) in one of the communiqués sent by Lord Hailsham to Prime Minister Harold Macmillan on September 10, 1962:

> Thank you for your minute of 3rd September on Ariel unfortunately damaged by Caliban as he girdled the Earth. We have got a great deal out of him during his life (short, but neither nasty nor brutish), in the shape of a large number of data, not yet fully analysed but said to be important. Although badly wounded in his solar paddles (the organ which recharges his ability to speak) he is not quite dead. He still utters intermittently – sometimes intelligibly – and is listened to by our monitors with respect.

JOINING THE CLUB

Following the passing of Ariel-1 and Prospero, British spaceflight efforts went into quiescent mode for a while. Then, in 1975, Britain became one of the founding states of the European Space Agency (ESA), which had emerged from the ELDO and the European Space Research Organisation (ESRO). This milestone was followed by the formation of the British National Space Centre (BNSC) in 1985 which, at the time, was the third largest contributor to ESA's general budget. But despite committing a significant amount of funding to space, the British government decided against contributing funds to the ISS because it thought such an investment did not represent value for money.

The years that followed Britain joining ESA were blighted by the myopic vision – at least when it came to spaceflight – of the Thatcher era. The Minister for the Department of Trade and Industry, Kenneth Clarke, labelled ESA as 'a hugely expensive club', an indication of the waning interest of the British government in anything related to spaceflight. It didn't take long before Britain's contribution to ESA's flagship projects – the Columbus space laboratory, the Ariane 5 launch vehicle and the Hermes spacecraft – dropped off and the BNSC's budget was put on ice. In the years that followed, the UK did make some contributions to the world of spaceflight, most notably the Beagle 2 probe (Figure 1.7) that was a part of ESA's Mars Express program. Sadly, Beagle 2 failed shortly after landing when its solar arrays failed to deploy. There was also Project Juno, but since that was financed by the Russians, it can hardly be classified as a British project.

A NEW SPACE PROGRAM?

Then, in 2009, the catalyst for the resurrection of Britain's manned spaceflight program came with the selection of Major Tim Peake to the European Astronaut Corps. The following year, the government established the UK Space Agency (UKSA) to replace the BNSC. At the time of the formation of the UKSA, the British public were still largely oblivious to the goings on in manned spaceflight. While Peake's selection caused a media buzz, the man from Chichester was still in astronaut training (Figure 1.8) and had not been assigned a mission. He graduated from training in 2010, but his mission assignment was not

Figure 1.7: Mars Express and Beagle 2. Beagle 2 was a Mars lander that was ferried to the Red Planet by ESA's 2003 Mars Express mission. If it had survived the landing, it would have searched for evidence of past life. Unfortunately, while the spacecraft deployed from Mars Express without a hitch, contact was lost before landing, and ESA declared the mission a bust. The fate of Beagle 2 remained a mystery until January 2015, when it was located intact thanks to images transmitted by the Mars Reconnaissance Orbiter. Careful scrutiny of the images revealed the lander to be intact, but with two of the four solar panels blocking the communications antenna. Credit: ESA

announced until 2013, and it was only then that the British public woke up to the fact that a British-sponsored astronaut was finally going to fly a mission.

"The space industry takes peaks and troughs, and in some respects the six new astronauts joined in something of a trough, with the cancellation of the U.S. Constellation program and the retirement of the shuttles. But it is like everything: you look to the future. Commercial transportation is a very exciting venture; there is the potential for Soyuz production to increase, and for the life expectancy of the ISS to

Figure 1.8:   Training at the Johnson Space Center in Houston, Texas. Credit ESA

be increased. The situation is quite optimistic. People are now looking not just to ISS but to the next step in the 2020s.

> *Tim Peake, being interviewed by the BBC's Jonathan Amos*
> *on 22 November 2010*

Two years later, Tim Peake found himself urinating on the rear right wheel of the coach that would take him into the heart of Baikonur to board a rocket that would take him to the ISS. The rest, as they say, is history. And for Britain's first official, government-sponsored astronaut, there was *a lot* of history.

REFERENCE

1. Ariel-1, the first international satellite project summary (NASA-SP-43). Baumann, R. C.

2

You mean we have a space program?

Figure 2.0: Credit: ESA

Message to President Reagan (after the successful first flight of the Space Shuttle)

Document type: Public Statement
Document kind: Message
Venue: No. 10 Downing Street
Source: Thatcher Archive

© Springer International Publishing AG 2017
E. Seedhouse, *Tim Peake and Britain's Road To Space*, Springer Praxis Books,
DOI 10.1007/978-3-319-57907-8_2

Editorial comments: Dispatched at 19:08 GMT
Importance ranking: Minor
Word count: 81
Themes: Foreign policy (USA)

I was thrilled to hear that the flight of the Columbia has been successfully completed.

The development of the space shuttle, which opens a new era in the space age, has been a great achievement and I share with you the pride you must now be feeling. Please convey to everyone involved — from the astronauts themselves to all the teams supporting them on the ground — my warmest congratulations on behalf of the British people. Margaret Thatcher, Prime Minister.

That little soundbite was sent by Prime Minister Margaret Thatcher on April 14, 1981, shortly after the successful landing of the Shuttle's inaugural mission. While that missive may have been supportive, Thatcher's interest in manned space flight was virtually non-existent. In fact, it was the Thatcher government that pulled the plug on the possibility of Britain joining the U.S., Canada, France, Germany, and … well, the rest of the world, by flying its own astronaut. Instead, thanks to Thatcher's myopic view of manned spaceflight, Britain was forced to watch non-government astronauts fly missions to *Mir* and on the Space Shuttle. Despite the political viscosity that defined the Thatcher era and beyond, British-born astronauts Michael Foale, Piers Sellers, Nick Patrick, Richard Garriott and Helen Sharman would subsequently add their names to the list of those who had flown in space, albeit on other countries' programs. Meanwhile, Britain continued its meager contribution to the European Space Agency (ESA). While Thatcher's enterprise-friendly macroeconomic policies allowed the British satellite market to thrive, her government refused to support a UK manned space program. So how did Major Tim Peake get selected as an ESA astronaut? To answer this, we need to go back to the Reagan Years.

THE REAGAN YEARS

In the 1980s, President Reagan invited ESA to join the effort to construct what would become the International Space Station (ISS). A decision was also made to invite the Russians so they could put their missiles to alternative use. But although Britain was a member of ESA, the government opted out of participation in ISS, complaining about lack of funds. This decision resulted in the widespread but unfounded British belief, at least among the decision makers, that manned spaceflight is an unaffordable luxury, whereas it is in fact a key part of any modern nation. To illustrate the point, by the time Helen Sharman flew her mission to ISS, no less than 21 other countries had already flown at least one astronaut (see Table 2.1). Abdicating from ISS was a shameful decision because Britain had been one of the first countries in space (as mentioned earlier, the country's first satellite was launched in 1962). In 2017, the nation still retains the unwelcome legacy of being

Table 2.1 Astronauts and Cosmonauts Timeline by Nationality

No.	Country	Name	Flight	Date
1	Soviet Union	Yuri Gagarin	Vostok 1	1961 Apr 12
2	United States	Alan Shepard	MR-3	1961 May 5
3	Czechoslovakia	Vladimír Remek	Soyuz 28	1978 Mar 2
4	Poland	Mirosław Hermaszewski	Soyuz 30	1978 Jun 27
5	East Germany	Sigmund Jähn	Soyuz 31	1978 Aug 26
6	Bulgaria	Georgi Ivanov	Soyuz 33	1979 Apr 10
7	Hungary	Bertalan Farkas	Soyuz 36	1980 May 26
8	Vietnam	Phạm Tuân	Soyuz 37	1980 Jul 23
9	Cuba	Arnaldo Tamayo Méndez	Soyuz 38	1980 Sep 18
10	Mongolia	Jügderdemidiin Gürragchaa	Soyuz 39	1981 Mar 22
11	Romania	Dumitru Prunariu	Soyuz 40	1981 May 14
12	France	Jean-Loup Chrétien	Soyuz T-6	1982 Jun 24
13	West Germany	Ulf Merbold	STS-9	1983 Nov 28
14	India	Rakesh Sharma	Soyuz T-11	1984 Apr 3
15	Canada	Marc Garneau	STS-41-G	1984 Oct 5
16	Saudi Arabia	Sultan al-Saud	STS-51-G	1985 Jun 17
17	Netherlands	Wubbo Ockels	STS-61-A	1985 Oct 30
18	Mexico	Rodolfo Neri Vela	STS-61-B	1985 Nov 26
19	Syria	Muhammed Faris	Soyuz TM-3	1987 Jul 22
20	Afghanistan	Abdul Ahad Mohmand	Soyuz TM-6	1988 Aug 29
21	Japan	Toyohiro Akiyama	Soyuz TM-11	1990 Dec 2
22	United Kingdom	Helen Sharman	Soyuz TM-12	1991 May 18
23	Austria	Franz Vlehböck	Soyuz TM-13	1991 Oct 2
24	Russia	Aleksandr Kaleri Aleksandr Viktorenko	Soyuz TM-14	1992 Mar 17
25	Belgium	Dirk Frimout	STS-45	1992 Mar 24
26	Italy	Franco Malerba	STS-46	1992 Jul 31
27	Switzerland	Claude Nicollier	STS-46	1992 Jul 31
28	Ukraine	Leonid Kadenyuk	STS-87	1997 Nov 19
29	Spain	Pedro Duque	STS-95	1998 Oct 29
30	Slovakia	Ivan Bella	Soyuz TM-29	1999 Feb 20
31	South Africa	Mark Shuttleworth	Soyuz TM-34	2002 Apr 25
32	Israel	Ilan Ramon	STS-107	2003 Jan 16
33	China	Yang Liwei	Shenzhou 5	2003 Oct 15
34	Brazil	Marcos Pontes	Soyuz TMA-8	2006 Mar 30
35	Iran	Anousheh Ansari	Soyuz TMA-9	2006 Sep 18
36	Sweden	Christer Fuglesang	STS-116	2006 Dec 10

the only one to have ever renounced its space capability. For decades, Britain not only gave up on manned spaceflight but also abandoned its launcher capability, opting out of helping ESA develop its Ariane launcher (Figure 2.1) and leaving the promising Reaction Engines *Skylon* program on life support. That left only satellites, although the British became very good at building these. But because the country had given up its launcher capability, Britain was reduced to hitching a ride on someone else's launcher, which in turn meant that companies building these satellites were driven to build small ones.

Figure 2.1: Ariane 5 flight V198. Ariane 5 is ESA's heavy lift launch vehicle that is part of the Ariane expendable launch system, capable of delivering payloads into geostationary transfer orbit or low Earth orbit. Operated and marketed by Arianespace, the rockets are launched from the Guiana Space Center in French Guiana. Originally designed to launch ESA's Hermes spaceplane, Ariane 5 was repurposed to launch satellites after the spaceplane idea was dropped in 1992. Credit ESA

Table 2.2. Major Tim Peake's Career Timeline

1990	Completed Combined Cadet Force as a Cadet Warrant Officer
1991	Member of a six-month Operation *Raleigh* expedition to Alaska
1992	Graduated as an Army Air Corps Officer. Served on attachments with the Royal Green Jackets as Platoon Commander
1994	Awarded flying wings. Served as a reconnaissance pilot and flight commander in Germany. Qualified as a combat survival instructor
1998	Qualified as a helicopter flying instructor
1999	Selected for an exchange posting with the U.S. Army
2002	Qualified as an Apache helicopter instructor
2005	Selected for test pilot training
2006	Served with the Rotary Wing Test Squadron at Boscombe Down
2009	After logging more than 3000 hours flying time on more than 30 types of helicopters and fixed wing aircraft, he retires from the army and is selected as an astronaut

Of course, none of this 1980s space politicking was on the radar for Timothy Nigel Peake. A student at Chichester High School in West Sussex, where he was described as the "calm, sensible type" (a phrase that one primary school report used to describe him at age 10), Peake was more interested in doing what kids at that age do. In Peake's case, much of his spare time was spent outdoors, engaging in camping, canoeing and cycling. Space wasn't even on his agenda, although family visits to the Science Museum piqued his interest. By the age of 15, Peake's interests revolved mainly around helicopters, and flying them. This was despite having a fear of heights, which he conquered by taking up rock climbing. While at school, he signed up with the cadet flying club and spent weekends flying gliders and tandem-seat Chipmunks. With his mind set on a career that involved flying, Peake studied hard to ensure he gained the qualifications necessary to apply to the military. As with most things Peake puts his mind to, the military venture was a success, and he graduated from the Royal Military Academy in Sandhurst in August 1992, as an officer in the British Army Air Corps. His first assignment was as platoon commander with the Royal Green Jackets, after which he was promoted to lieutenant. Three years later, he was promoted to captain and qualified as a helicopter instructor after graduating from the Defence Helicopter Flying School at RAF Shawbury in 1998. He then completed tours in Northern Ireland (flying Gazelle helicopters), Afghanistan, Bosnia and Kazakhstan (home to the Baikonur Cosmodrome). Another promotion followed in 2004 when Peake achieved the rank of Major and that same year he also graduated from the Empire Test Pilots School, having been awarded the Westland Trophy for best rotary wing pilot. In 2006, Peake was awarded the Commander-in-Chief's Certificate for Meritorious Service, for exemplary service to the British Army. That year, he also completed his educational achievements by attaining his bachelor of science degree (Hons) in Flight Dynamics and Evaluation from the University of Portsmouth.

And then there was that astronaut job.

"Tim has done so many things in his life that when he told us he had been chosen, it didn't come as a surprise. Tim has always shown huge determination to succeed. I can remember school reports from Chichester High which said: "Tim never gives up.""

Nigel Peake, Tim's father

"He persevered with swimming, although it didn't come naturally. He wasn't that good at it. Even if he was the last to finish, he'd keep plugging away."

Mrs. Angela Peake, Tim's mother

If there is one quality every astronaut has in abundance, it is that ability to keep on 'plugging away'. This is probably not that surprising, when you consider just how difficult it is to be selected as an astronaut in the first place. But that's just the beginning. Training is on another level entirely. First, there is the small matter of 18 months of basic training and then, once selected for a flight, astronauts spend another 18 months training just for that mission. Given the broad spectrum of tasks that astronauts must be proficient in, it is also probably not surprising that, every once in a while, an astronaut has to cope with a task they find a little more challenging than anticipated. In Peake's case, the hurdle that required him to apply his 'plugging away' skills was mastering the Russian language.

THE CATALYST – DAVID WILLETTS

From the Thatcher era onwards, Britain missed out on an awful lot of space activity, but in 2012 there was an opportunity to end the stagnation when ESA's ministerial meeting took place to establish budgets. The French and the Germans were at loggerheads and there was the risk that ESA might fall short of its ISS commitments. Fortunately, the British minister responsible for space at the time – the Rt. Hon. David Willetts (the Minister of Universities and Science, Figure 2.2) – finally reversed decades of British government policy and offered a contribution that allowed Britain to become part of ESA's program for the ISS. There was one condition, however: that this contribution secured a flight for Tim Peake. And so, the deal was done. Thanks to the foresight of David Willetts (helped in part by the work done by Lord Drayson of the previous Labour government), Britain was finally in a position to become one of the world's leading space-faring nations.

Even now, however, despite all the positive spin generated by Tim Peake's flight, there are still those who argue that Britain should stay out of manned spaceflight (see sidebar). But there are arguably more reasons for the country to continue to support further flights than there are to return to the old days of just building satellites. For one thing, humans can do a lot that robots can't. Take the Apollo missions, for example. The astronauts travelled 17 miles on the lunar surface using the lunar rover. The Mars rover on the other hand, neat as it is, travelled just three miles in three years. Three years! And then there are the positive side-effects of having a British astronaut in space. After Peake's flight, there was a big increase in the number of students who wanted to study science and engineering, and that can be no bad thing with Britain's manufacturing sector constantly crying out for a technically competent workforce.

So, 25 years after Helen Sharman's flight, Peake finally put Britain back on the manned spaceflight map, and for many it was a flight that was long overdue. After all, this was a country that had produced international space celebrities such as Arthur C. Clarke, Colin Pillinger of Beagle 2 fame, and Stephen Hawking. Just no astronauts!

Figure 2.2: Science Minister David Willetts (right) with Tim Peake and museum director Ian Blatchford at the announcement of Peake's ISS mission. Credit: The Science Museum

Manned or Unmanned?

In 2007, a working group addressed the issue of Britain missing out by not having a manned space program. The group, which was set in motion by Ian Pearson (Minister of State for Science), recommended that a manned spaceflight program should be established by undertaking precursor missions to the ISS. That way, the group argued, Britain could gradually create a corps of four astronauts who could build experience and also inspire students at schools and colleges. The recommendations made by the group seemed sensible enough, but there were some hurdles that stood in their way. One of these was the fact that the Treasury exerted a vice-like grip on any public project, and was reluctant to hand over money for those projects that didn't have an immediate, or at least a near-term, spinoff for industry. And while space was deemed profitable, manned spaceflight was still judged to be a luxury, with the working group's case hardly helped by the $100 billion cost of the ISS. A second hurdle was the British robotic science lobby, which traditionally had used most of the meager pittance of Britain's space budget. That budget would be stretched beyond breaking point if some of the money had to be siphoned off to send British astronauts on a low Earth orbit joyride. It didn't matter that manned space-flight proponents like the British Interplanetary Society (BIS) had pointed out that most of the developed world seemed able to fly satellites *and* astronauts. Take Sweden, for example. At the time of the working group's recommendations, Sweden had already sent its astronaut, Christer Fuglesang, on his first space mission – STS-116 in 2006. Sweden was also operating a rocket range in Kiruna *and* it built satellites, and all on a shoestring budget of $100 million a year – less than half of Britain's budget at the time.

THE ISS AGA

The announcement of Tim Peake's launch was a huge deal for British involvement in space and the ISS program, with which the UK had always had a hot-and-cold relationship. The beginning of that relationship came on January 29, 1998, when the UK was one of 15 nations to sign the ISS Intergovernmental Agreement (IGA – see sidebar), which was the instrument that formally – and finally – brought the ISS (Figure 2.3) off the drawing board and into reality.

The ISS Intergovernmental Agreement (IGA)
The ISS is a program of collaboration between the U.S., Russia, Canada, Japan and 10 member states of Europe. The legal framework governing the development, utilization and operation of the orbiting outpost is outlined in the IGA in Article 1 as follows:
Article 1: Object and Scope
1. The object of this Agreement is to establish a long-term international cooperative framework among the Partners, on the basis of genuine partnership, for the detailed design, development, operation, and utilization of a permanently inhabited civil international Space Station for peaceful purposes, in accordance with international law. This civil international Space Station will enhance the scientific, technological, and commercial use of outer space. This Agreement specifically defines the civil international Space Station program and the nature of this partnership, including the respective rights and obligations of the Partners in this cooperation. This Agreement further provides for mechanisms and arrangements designed to ensure that its object is fulfilled.
2. The Partners will join their efforts, under the lead role of the United States for overall management and coordination, to create an integrated international Space Station. The United States and Russia, drawing on their extensive experience in human space flight, will produce elements which serve as the foundation for the international Space Station. The European Partner and Japan will produce elements that will significantly enhance the Space Station's capabilities. Canada's contribution will be an essential part of the Space Station. This Agreement lists in the Annex the elements to be provided by the Partners to form the international Space Station.

Figure 2.3:   The International Space Station. Credit: NASA

> 3. The permanently inhabited civil international Space Station (hereinafter "the Space Station") will be a multi-use facility in low-Earth orbit, with flight elements and Space Station-unique ground elements provided by all the Partners. By providing Space Station flight elements, each Partner acquires certain rights to use the Space Station and participates in its management in accordance with this Agreement, the MOUs, and implementing arrangements.
>
> 4. The Space Station is conceived as having an evolutionary character. The Partner States' rights and obligations regarding evolution shall be subject to specific provisions in accordance with Article 14.

The IGA also allowed the ISS Partner States to extend their national jurisdiction to space, which means that the modules provided by those states are incorporated into the territories of the Partner States. This means that each Partner State retains control over the modules it registers and also control over its personnel, as stated in Article 5 (see sidebar).

IGA Article 5: Registration, Jurisdiction and Control

1. In accordance with Article II of the Registration Convention, each Partner shall register as space objects the flight elements listed in the Annex which it provides, the European Partner having delegated this responsibility to ESA, acting in its name and on its behalf.

2. Pursuant to Article VIII of the Outer Space Treaty and Article II of the Registration Convention, each Partner shall retain jurisdiction and control over the elements it registers in accordance with paragraph 1 above and over personnel in or on the Space Station who are its nationals. The exercise of such jurisdiction and control shall be subject to any relevant provisions of this Agreement, the MOUs, and implementing arrangements, including relevant procedural mechanisms established therein.

But while the UK was a signatory to the ISS IGA, it was the only country that did not contribute funding to the space station program. That decision was in keeping with the UK's long-standing policy that prevented the country from putting up any funding for manned spaceflight. The decision also meant that Britain was left out when it came to sharing the benefits of the research that resulted from the science on the ISS. But, after decades in the wilderness – at least as far as manned spaceflight was concerned – Britain finally made a decision (fueled no doubt by the fact that the space industry contributed £9 billion a year to the British economy) to increase its involvement in the space arena. The first step was to establish the UK Space Agency (UKSA, Figure 2.4) in April 2010, an event that turned out to be a catalyst for a number of other benefits that would ultimately lead to Peake being offered his ISS flight opportunity. The creation of the UKSA (see sidebar), on the back of Peake's selection as an astronaut, sparked a significant

Figure 2.4: UK Space Agency. Also known as the UK Space Agency, or just plain UKSA, the organization is an executive agency of the British government and is responsible for the country's civil space program. Established on April 1, 2010, the UKSA replaced the British National Space Centre (BNSC) and assumed responsibility for budgets and government policy. Before the creation of the UKSA, the space industry in the UK was valued at around £6 billion and supported 68,000 jobs. The aim of the UKSA is to boost those numbers over 20 years to £40 billion and 100,000 jobs. Credit: UKSA

increase in lobbying by space advocacy groups. Now, all of a sudden, UK manned spaceflight was on a roll. And it just kept on getting better.

Just two years later, Britain contributed £1.2 billion to ESA (which made the country the third largest contributor), of which an unprecedented £12.4 million was earmarked for ESA's European Life and Physical Sciences (ELIPS) program. This contribution meant that the UK would now be granted access to the microgravity environment of the station to perform research. Another important contribution made by the UK was the £16 million invested in ESA's efforts to design and build the Automated Transfer Vehicle (ATV, Figure 2.7) Service Module (SM) destined for NASA's Orion crew vehicle. That contribution kick-started a number of construction contracts for British companies, especially in the areas of propulsion and communication. The funding also meant that, with Britain now being a financial contributor to the Orion SM venture, the country was promoted to the status of ISS partner nation (the British flag had quietly been removed from the ISS in 2010) – at least for the 2015 to 2020 timeframe.

The catalyst for all these fortuitous events was Peake's selection as an ESA astronaut in 2009. At the time of his selection, Peake was identified as a 'European astronaut of British origin' and not as a 'British astronaut'. But that changed in 2013 – just six months after Britain's financial contributions – when Peake was assigned to his ISS mission. That event was marked by some controversy, however, since the original assignment was for a short duration mission of just 10 days. The British took umbrage at that decision, pointing out that Danish astronaut Andreas Mogensen had been assigned to a long duration mission, despite the fact that Denmark's contribution was less than Britain's. After some political arm-wrestling, ESA reversed its decision and handed off the short duration to the Danes and the long duration slot to the Brits. Everyone was happy. Well, perhaps not Mogensen.

UKSA Space Strategy
"The UK will be a recognised and valued participant in human spaceflight and space environments research – in low Earth orbit, on analogue platforms and in deep space exploration. Advancing scientific knowledge and technological capabilities as a pathway to growth will positively augment the UK economy and provide measurable societal benefits in sectors such as healthcare, communications and education."

UK's first Space Environments and Human Spaceflight national strategy,
released by UKSA, July 2015

3

Right place, right time

Figure 3.0:   Credit: ESA

© Springer International Publishing AG 2017
E. Seedhouse, *Tim Peake and Britain's Road To Space*, Springer Praxis Books,
DOI 10.1007/978-3-319-57907-8_3

"We want to find high-calibre men and women in Europe to prepare to meet the challenges of ISS exploitation and human exploration of our solar system in the 21st century. As of May 2008, ESA will be searching in each of its 17 member states for the best candidates to make this vision a reality."

Michel Tognini, former astronaut and
chief of the European Astronaut Centre, May 2008

RECRUITMENT DRIVE

Europe's third astronaut hunt began on May 19, 2008, when the European Space Agency (ESA) announced that it was searching for four new members to join the astronaut corps. Since this astronaut recruitment campaign was just the third in 40 years, the competition for places was predicted to be fierce (see Appendix I for the details of the selection criteria), with ESA expecting that they would receive thousands of applications. Like all the budding space-farers that went before them, ESA's latest crop of astronauts were required to be highly skilled, able to maintain their composure under pressure, and work effectively as a vital cog in the complex machine that is the International Space Station (ISS). For most applicants keen to get into the business of being an astronaut, the year-long recruiting process was daunting enough, but if you happened to be British there was another barrier to face; namely that long-standing resistance the British government had to investing in manned spaceflight. While Britain's contribution to ESA had increased steadily over the years, its £208 million contribution in 2008 (which made it the fourth-largest contributor to ESA) was earmarked for robotic missions. And while ESA officials stressed that an applicant's passport would have no bearing on whether or not they were selected, it must have been difficult not to be mindful of the fact that historically, astronauts had been chosen from those countries who ploughed the most into manned spaceflight.

ESA's 2008 campaign (Figure 3.1) came at a time of flux in the manned spaceflight arena, because the Shuttle had been slated for retirement and NASA had recently announced plans (which were subsequently cancelled) to establish a lunar outpost. The year-long process began with an online application that was submitted by 8,413 candidates.

The application required candidates to agree to release data and sign a privacy statement, in addition to providing standard demographic data, details of their education and professional expertise and specific background information. Standard astronaut selection stuff, in other words. Candidates were also required to provide evidence of a current Class 2 JAR-FCL 3 medical (see Appendix I). When the dust had settled on the first round, there were 912 candidates left standing. Next on the menu was a psychology assessment that tested knowledge of engineering, technology, mathematics, logic and reasoning, spatial orientation and multi-tasking. Once the results of the psych test had been analyzed, the 912 candidates had been reduced to just 192. This group was required to complete a second psychology test that assessed problem-solving techniques,

Figure 3.1: Credit: ESA

examined behavior-oriented personality diagnostics, and included a Dyadic Cooperation Test[1] and structured interviews. The second psychology test reduced the candidates further to just 45, who would then face the medical examination. This would follow the standards outlined in the ISS Medical Volume A, and required candidates to complete a thorough medical history and questionnaire review, and submit to myriad clinical and specialist examinations (Table 3.1), anthropometry, imaging diagnostics, and a final review by the ESA Medical Selection Board (Table 3.2).

Of the 45 remaining applicants, just 22 passed the medical examination. Usually, the main reason for disqualification of a candidate is due to problems with their vision, but in the 2008 ESA campaign the primary cause for disqualification was due to cardiovascular disease, or CVD.

---

[1] The Dyadic Cooperation Test assessed traffic management, decision-making, reliability, cooperation, working style, and stress resistance.

Table 3.1 Types of Medical Exams

- MRI: Magnetic Resonance Imagery
- EBCT/MDCT: Multi-Detector Computer Tomography (Coronary Calcium Scores)
- Carotid Doppler and ECHO
- CT: Computerized Tomography / CAT: Computerized Axial Tomography Scan

Table 3.2 Results of ESA's 2008 Astronaut Medical Examination

Cause of disqualifications	Number of applicants disqualified	Percentage of applicants	Percentage of disqualified applicants
Cardiovascular	14	31	61
Vision	7	16	30
Thyroid	2	4	9
Ear Nose Throat	1	2	4
Respiratory	1	2	4
Gastrointestinal	1	2	4
Genitourinary	1	2	4
Neurological	1	2	4

For those living in the U.S. the answer to the question, 'How do I become an astronaut', has always been simple: Apply to NASA (see sidebar). But for applicants from Britain, the options had been much more restricted, until the 2008 ESA selection. One option was to have dual nationality, which is how Michael Foale got into the NASA program during the Shuttle era (see Chapter 4). Another option was to apply to ESA, but although they were not officially barred from the program, British applicants had next to no chance of succeeding because of the government's continued refusal to invest in manned spaceflight.

NASA's Astronaut Selection Timeline

Sep 2007	Vacancy Announcement opens
Jul 1, 2008	Vacancy Announcement closes
	3564 applications received
Sep 2008	Applications reviewed to determine Qualified Applicants
	Over 700 disqualified, leaving 2,800
Oct 2008	Qualified Applications
	Reviewed to determine Highly Qualified applicants
	~ 450 selected
	References for Highly Qualified applicants contacted via
	mail, and background checks done
Nov 2008	Highly Qualified applications reviewed to determine
	Interviewees:
	~120 selected
Nov 2008 –	Interviewees brought to JSC for preliminary interview,
Jan 2009	medical evaluation and orientation.

	(groups of 10 invited to Houston for 2.5 days of medical testing and interviews)
Feb 2009	Finalists determined: ~40 selected
Feb-Mar 2009	Finalists brought to JSC for additional interview and complete medical evaluation.
May 2009	Astronaut Candidate Class of 2009 announced
Aug 2009	Astronaut Candidate Class of 2009 reports to the JSC

A third option was the private route, which is how Helen Sharman launched her spacefaring career. Sharman's eight-day mission on board the aging Russian station *Mir* in May 1991 turned the 27-year-old scientist into a national celebrity. The Prime Minister welcomed her to Downing Street, streets were named after her and, in her home town of Sheffield, a star was put in the sidewalk. For years after her mission, she travelled the country talking about her experiences. And then she pulled a conjuring trick that David Copperfield would have been proud of. She simply disappeared. Well, not literally. Today, Sharman (Figure 3.2), an intensely private individual, works in London's Imperial College as operations manager for the chemistry department. After being thrust into the limelight following her mission to *Mir*, Sharman eventually got fed up of being asked how to go to the toilet in space, and retreated into a self-imposed anonymity (few of her students are aware that she is one of only 60 female astronauts). Sharman would probably have stayed out of the spotlight, had it not been for the UK Space Agency's attempt at rewriting history. In 2013, shortly after Major Peake had been selected for the Principia mission, the UKSA announced that Britain would soon be flying its first official astronaut, without explaining what 'official' meant.

"ASTRONAUT WANTED, NO EXPERIENCE NECESSARY"

That was the strapline advertising Project Juno in the late 1980s, a program designed partly to boost UK-Russian relations. Sharman's journey to *Mir* started when she heard the above announcement on the radio, while working as a research chemist in 1989. She was one of 13,000 hopefuls who applied. The *Astronaut Wanted, No Experience Necessary* line was typical of the understated humor of a country whose imperial history, and technological capabilities in the 20th century, was at odds with its dearth of support for manned spaceflight. The (brief) glory days of Britain being at the sharp end of space exploration, with its Prospero satellite, had long since evaporated thanks to the Thatcher government, which had successfully gutted even a sliver of hope of a manned space program. Despite this, the pre-*Challenger* era did witness efforts to send a Briton into orbit. Squadron Leader Nigel Wood had been selected to fly with NASA in June 1986, as a Payload Specialist in support of Britain's Skynet project. Wood completed his astronaut and payload training, but when *Challenger* was destroyed in January 1986, Wood's chances of reaching space came to end.

Three years later, research chemist Helen Sharman was sitting in traffic on her way home from her job at the Mars Confectionary company in Slough, Berkshire, when she heard the

Figure 3.2: Helen Sharman. For eight years following her mission to *Mir*, Sharman was self-employed; communicating science to the public, promoting her book *Seize the Moment*, and appearing on various radio and television shows. But after being asked the same questions over and over again, she retreated from the public eye. Today, she works as Operations Manager for the Department of Chemistry at Imperial College, London, although she continues space-related outreach activities. As Britain's first manned spaceflight celebrity, Sharman has received a number of awards: In 1991, she was chosen to light the flame at the Summer Universiade, held in her home town of Sheffield; she was awarded the OBE in 1993; Wallington High School of Girls named a house after her, as did a grammar school in Sutton; the science block – named Sharman House – of Buller's Wood School was opened by Sharman in 1994; she received the title of Honorary Fellow from Sheffield Hallam University in 1991, an Honorary Doctor of Science degree from the University of Kent in 1995, an Honorary Doctor of Technology degree from the University of Plymouth in 1996, an Honorary Doctor of Science degree from Southampton Solent University in 1997, a third Honorary Doctor of Science degree from Staffordshire University in 1998, and a fourth from the University of Exeter in 1999. Her fifth Honorary DSc came from Brunel University in 2010. Credit: Russian Space Agency

radio announcement. As she noted in her autobiography, *Seize the Moment*, the events of that day would become "the crucial, pivotal moment in my life." Two years later, Sharman became the first Briton in space, and in doing so meant that Britain became one of only three nations to have a woman as its first astronaut. A native of Sheffield, Yorkshire, Sharman was born on May 30, 1963. Graduating from the University of Sheffield in 1984 with a degree in chemistry, Sharman first worked as an engineer in London, before pursuing the PhD in chemistry that led to her employment with Mars Confectionary. When she heard the astronaut announcement, Sharman realized that she checked all the boxes. She was British, aged between 21 and 40 years old, with a scientific background and proven language capability, since she already spoke French and German. She duly applied, and when the list of 13,000 people had been whittled down to 150 a few weeks later, Sharman was still in the running. After the medical evaluations, she was one of the final 32 candidates, all of whom were summoned to a meeting at Brunel University in Uxbridge in August 1989. It was there that she met Major Tim Mace, a Royal Army Air Corps helicopter pilot and sky-diver. In *Seize the Moment*, Sharman reckoned that Mace was the lead candidate for the flight.

> "The stakes were much higher than any of us had realised. The publicity would fuel the sponsorship, but for most of the applicants it would inevitably mean rejection in public."

> *Excerpt from* Seize the Moment: Autobiography of Britain's First Astronaut,
> *Helen Sharman, with Christopher Priest*
> *(London: Gollancz, 1993 - ISBN 0-575-05819-6)*

After the meeting at Brunel, the candidate list was winnowed down to 22, and then to 16. It was at this stage that Sharman figured her run had come to an end, because the Russians expressed their desire that the two finalists be of the same gender, although they didn't specify male or female. Since there were only two females left in the running, Sharman reckoned her odds of being selected were slim to none. The 16 was soon reduced to just 10 and now Sharman found herself to be the only female left. Then, following a visit from Soviet flight surgeons, just six candidates remained. In November 1989, the final four were announced, and Sharman and Mace had both made it, along with Royal Navy flight surgeon Gordon Brooks and aeronautical engineer Clive Smith. A few weeks later, Sharman and Mace were selected for the mission – dubbed Project Juno – and flew to Star City to begin 18 months of mission training.

Shortly after the two British candidates reported for cosmonaut training, Project Juno (Figure 3.3) hit a roadblock. The Soviets had charged the British $12 million for the flight,

Figure 3.3: Project Juno. Public domain

but by early 1990 it was apparent there was a shortfall in funding that threatened the mission with cancellation. Corporate sponsors included Memorex, Interflora and British Aerospace, but the money put up by these companies amounted to only $1.7 million. As expected, the British government refused to support the mission. So much for furthering international relations! The revenue drama dragged on for the whole of 1990, and it wasn't until December that Sharman and Mace got word that Moscow Norodny Bank (thanks to encouragement from Premier Mikhail Gorbachev, if rumors are to be believed) had agreed to cover the costs. Sharman described the bank as "our white knight." Score one for the Russians. For Britain, the revenue debacle was one that shed a poor light on the British government. In fact, it was another damning and pitiful indictment of the myopic and narrow-minded vision the Thatcher government had when it came to manned spaceflight matters. A downright embarrassment. And it wasn't as if Britain couldn't afford it. Take Germany, for example. This country had been struggling with the obscene cost of reunification after the collapse of the Berlin Wall, but the country still sank $25 million into a mission to fly its cosmonaut to *Mir*. Britain's cosmonaut meanwhile? Well, Sharman's mission was the equivalent of a hitchhike.

Meanwhile it was time to select the prime crew, a decision that was finalized on February 19, 1991, when Air Vice Marshal Peter Howard, an aerospace physician who had been placed in charge of Project Juno's astronaut selection, informed Sharman that she would be flying to *Mir*. Sharman and Mace continued to train together until their formal training concluded at the end of April 1991. There was less than a month before the launch of Soyuz TM-12. Flying with Sharman to the aging Russian outpost would be Commander Anatoli Pavlovich Artsebarski, and Flight Engineer Sergei Krikalev.

Artsebarski's background had included two years preparing the Buran shuttle, while Krikalev was about to make the second flight of what would be an illustrious six-flight career (in 2017 he is still third in the all-time list for the most cumulative time in space: 803 days). On May 18, 1991, the prime crew together with their backups (Alexander Kaleri, Tim Mace, and Alexander Volkov) prepared for launch at Tyuratam. On the launch pad, adorned with a Soviet Hammer and Sickle and a British Union Jack, the Soyuz TM-12 stood waiting for the crew. At 3:50 p.m. Moscow Time, TM-12 thundered into orbit.

> "It did not happen gradually. One moment, it was burning ferociously behind me, in the next it stopped completely. One moment I was being pressed hard into my seat and in the next I was not. I had been straining against the G-force without realising I had been doing so."
>
> *Excerpt from* Seize the Moment

Five minutes after launch, the central core stage was jettisoned. Three minutes after the separation of the third stage, the crew marked their arrival in space. Two days later, following an uneventful transit to *Mir*, the Soyuz TM-12 made its final approach to the station. Everything was looking good until the spacecraft closed within 200 kilometers of its target. That's when the KURS rendezvous system malfunctioned, forcing Artsebarski to perform the final approach under manual control. After docking, the visitors were treated to bear hugs and the customary bread-and-salt welcome from Viktor Afanasyev and Musa Manarov, the resident crew. Since the British commercial participation in the mission had been significantly downgraded, there were very few British experiments to keep Sharman busy. In fact, the content of her experiment program was almost exclusively Soviet, with a focus on life sciences. She tracked her heart rate, checked her reaction time, took blood samples and helped out with monitoring the life support system by taking air samples. She also helped grow wheat seedlings and potato roots, and found time to take plenty of photos of Britain.

The fact that Sharman's mission was Soviet-orchestrated and Soviet-financed did not escape the scorn of some British journalists, notably Tim Furniss, who wrote in *Flight International*: "Sharman will operate 17 biotechnological, medical, and technical experiments – all Soviet ... the British element of the Project Juno mission will be Sharman herself ..." That said, most of the coverage of Sharman's historic mission was positive. And, as with all missions to *Mir*, day-to-day life on board the orbiting outpost was nothing if not eventful. Shortly after docking, Sharman was treated to the first of what would be several station malfunctions, when *Mir* (Figure 3.4) suffered problems orienting its solar arrays. One side-effect of this was that the level of background noise in the station plummeted, as the fans and circulating pumps shut down. From being a noisy place with a background decibel level of around 70, *Mir* suddenly went quiet ... and then dark as the lights switched off, leaving the crew with just a single emergency fluorescent tube. The crew had to wait until the station re-entered sunlight, allowing them to adjust the

Figure 3.4: Space station *Mir*. Credit: NASA

orientation of the arrays to allow the batteries to be recharged. To old hands Afanasyev and Manarov, it was business as usual, and they reassured Sharman that this sort of stuff happened all the time. Which it did.

As was the standard for international missions in those days, Project Juno lasted a week (7 days and 21 hours), and on May 26, at 09:13 Moscow Time, Soyuz TM-11 made tracks for home by undocking from the Kvant-1 module. Sharman, as with other short-duration visiting crewmembers, would return to Earth on the older Soyuz with the outgoing resident crew, Afanasyev and Manarov, who had been replaced aboard *Mir* by Artsebarski, Krikalev and their new vehicle. A retrofire burn was performed at 12:38 p.m., and eleven minutes later the drogue parachutes deployed. Fifteen meters above ground, retrorockets fired to cushion the impact, and Britain's first manned spaceflight came to a close. It was 13:03 Moscow Time. Having gotten a taste for the astronaut business, it was no surprise that Sharman (and Mace) applied to join ESA's astronaut corps in 1992 and 1998, but without success. In the years following Sharman's mission there were other Britons who flew in space, but not with the support of the British government. Piers Sellers, Nick Patrick and Michael Foale were all born in the U.K, but had to take advantage of acquired U.S. citizenship to fly via the NASA route. Other dual nationality Britons such as Richard Garriott and Mark Shuttleworth took the commercial route and flew as paying spaceflight participants.

> "It's about understanding the universe and how, one day, humans can survive off this planet. Britain should be helping with that research. It's expensive, but it's important. And it's good value. Space missions capture the imagination. It turns people on to science."
>
> *Helen Sharman, speaking about the need for increased UK space funding.*

Anyone with even the flimsiest knowledge of the British government's persistent policy of opting out of manned spaceflight can't have been too surprised that it took two decades after the *Astronaut Wanted, No Experience Necessary* announcement before another British spaceflight opportunity presented itself. That was when Tim Peake got the call.

> "I just wanted to see how far I could get. Being an astronaut wasn't something I ever envisaged doing, I loved flying and was happy in my career but then ESA applications opened up so I thought I'd give it a go!"
>
> *Tim Peake, on his decision to apply to ESA*

It was May 18, 2009, and ESA astronaut applicant Tim Peake was still living in Army housing in Larkhill, Wiltshire. He had just retired from the Army Air Corps after 18 years of flying helicopters and the future looked promising, as he had landed a job as a test pilot with Agusta Westland Helicopters in Somerset. He had also just finished the year-long astronaut selection process for ESA's Astronaut Corps. Like most Britons, Peake had never seriously considered the job of astronaut as a viable career path because he was missing one vital qualification: an American passport. But, in the spirit of 'nothing ventured, nothing gained', Peake jumped at the chance when the call was announced in 2008.

It wasn't long before he reckoned he had a pretty good chance of being selected, especially since many of the tests were very similar to those used by the military to select their pilots: hours of computer-based tests, cognitive skills, spatial awareness tests and endless psychological questionnaires. Another similar selection criterion was the medical exam, although the astronaut version was more invasive (it included a double enema and sigmoidoscopy). And when the dust had settled on all the tests, Peake found himself as one of the final 10 candidates that were invited to meet ESA's Director General, Jean-Jacques Dordain. That meeting had been a month earlier, and on the evening of May 18 Peake was trying to relax with a glass of wine, wondering whether he had made the cut. Then the phone rang, and the voice at the other end asked the retired Army Air Corps pilot if he might be interested in joining the European Astronaut Corps. Peake was on a flight to Paris the next day.

AFTER SELECTION, TRAINING

"A human being should be able to change a diaper, plan an invasion, butcher a hog, conn a ship, design a building, write a sonnet, balance accounts, build a wall, set a bone, comfort the dying, take orders, give orders, cooperate, act alone, solve equations, analyze a new problem, pitch manure, program a computer, cook a tasty meal, fight efficiently, die gallantly. Specialization is for insects."

Robert A. Heinlein. 'Time Enough for Love'

While an astronaut doesn't need to be proficient in all the tasks itemized by Heinlein, versatility *is* a key quality demanded by space agencies when recruiting astronauts. NASA, the CSA and ESA search for versatile individuals who have high technical competence and the ability to work in a team. The selection team also looks for a hybrid: a researcher who can handle the science and operate manipulators (Figure 3.5) such as the Canadarm2, and who also has the physical ability to endure spacewalks that may last six hours or longer. Given these requirements, it is tough for applicants to meet the grade and show they possess all those qualities. But the myriad qualifications are necessary, since training an astronaut is a considerable investment for any agency and the support needed both before and during a space mission is costly. The average cost per crewmember per day on board the ISS is a whopping $7.5 million. *Per day!* It also takes years to plan a space mission and hundreds of people to prepare the astronauts and the spacecraft, and since astronauts are pivotal to the success of a mission and flight opportunities are so limited, space agencies obviously want to ensure that those selected will make the best possible use of the very, *very* precious time they will spend in space.

Of the few agency-sponsored astronaut training programs around the world, ESA's is perhaps not only the most challenging to follow, but also the most challenging to organize. This is because the decentralized training provided by the international partners that participate in the ISS program demands a high degree of coordination and synchronization. Furthermore, the teaching methods, training concepts *and* the multicultural backgrounds of the astronaut candidates and their instructors must be standardized. Once the training

Figure 3.5: Astronaut selection. Major Peake at the rover control workstation in the International Space Station's Columbus science module. Credit: ESA

has been coordinated and synchronized, the training aspects must then fit into one seamlessly integrated schedule. This is a challenge in itself, since each astronaut has an individually tailored training plan, meaning no two astronauts get exactly the same training at the same time. If you consider that between 30 and 40 astronauts and cosmonauts undergo training in one year at five different sites, you get some idea of the tremendous organizational effort required. Fortunately, the ISS partner agencies have successfully managed to synchronize the training for more than a decade now.

Phases of astronaut training

Like every ESA astronaut before him, Peake started his astronaut training cycle by first completing the 16-month Basic Training increment at the European Astronaut Centre (EAC) in Cologne, Germany (Figure 3.6). This phase, which provided Peake with a solid introduction to his future career as an astronaut, comprised four training blocks:

1. Introduction.
2. Fundamentals.
3. Space Systems and Operations.
4. Special Skills.

Figure 3.6:   The European Astronaut Center. Established in 1990, the EAC is located near Cologne, Germany. The EAC comprises a staff of more than 100, many of whom are employees of the German Aerospace Center (DLR) and CNES, the French Space Agency. The EAC is ESA's center of excellence when it comes to astronaut training and selection, medical training and support, and the scheduling of an astronaut's tasks and flight assignments in preparation for missions to the International Space Station. The EAC is also ground zero for training astronauts on European-built ISS hardware, such as ESA's Columbus laboratory. At the time of its founding in 1990, there were just three astronauts based at EAC: Ulf Merbold (Germany), who flew on the first Spacelab mission in 1983, Wubbo Ockels (Netherlands), who flew on the Spacelab D1 mission in 1985, and Claude Nicollier (Switzerland) who flew on STS-46, STS-61, STS-75 and STS-103 between 1992 and 1999. In 1998, ESA member states combined the existing national astronaut teams and the European Astronaut Corps came into being. Credit: ESA

1. Introduction

This phase is a little like the first day at work, since it provides the new hires a broad overview of the arena in which they will be working for the next ten to twenty years. For example, they are introduced to the policies of the major space-faring nations, their space agencies (with special emphasis on ESA) and the major manned and unmanned space programs. Spicing up this phase are lectures covering the basics of space law and the intergovernmental agreements governing worldwide cooperation in space.

2. Fundamentals

The second phase of basic training provides the candidates with basic knowledge of various technical and scientific disciplines. The objective of this phase is to ensure that all new astronaut candidates (or 'ascans'), from whatever their diverse professional backgrounds and expertise, have a common minimum knowledge base in subjects relevant to their new career. Although the phase deals only with the fundamentals, an astronaut's job requires knowledge of so many different subjects that candidates often feel as if they're 'drinking from a fire-hose' as they try to take in all the information fed to them by their instructors. In addition to covering technical disciplines such as spaceflight engineering, electrical engineering, aerodynamics, propulsion and orbital mechanics, this phase also includes an introduction to science disciplines such as research into weightlessness (in human physiology, biology and material sciences), Earth observation and astronomy.

3. Space Systems and Operations

This phase provides candidates with a detailed overview of all the ISS systems, such as guidance navigation and control (GNC), thermal control, electrical power generation and distribution, command and tracking, and life support systems. It also introduces the ascans to robotic, EVA and payload systems, in addition to the major systems of those spacecraft which service the ISS, such as the Russian Soyuz (Figure 3.7) and Progress spacecraft and SpaceX's Dragon (Figure 3.8). Additionally, candidates become acquainted with ground systems such as development and test sites, launch sites and training and control centers.

4. Special Skills

The Special Skills phase focuses on developing the skills required for generic robotic operations, rendezvous and docking, Russian language, human behavior and performance, and scuba-diving. Scuba-diving (Figure 3.9) is an essential skill for those astronauts who will eventually perform EVAs. The Special Skills phase completes the basic training of ESA astronaut candidates and marks the beginning of the advanced training increment.

Advanced Training

During the advanced training phase, the ascans learn how to service and operate the different ISS modules, systems and subsystems, how to fly and dock transport vehicles like the Russian Soyuz vessel, and how to survive if things go pear-shaped (Figure 3.10). They also learn how to perform scientific experiments in ESA's ISS research module, Columbus (see Panel 3.1), and how to use the equipment fitted in the module's payload racks, such as the European Physiology Modules Facility (EPM).

Advanced training takes a year to complete. The training is conducted in international astronaut classes and includes training units at all ISS partner training sites. These training sites are located at Houston in the U.S., Star City near Moscow (Russia), Tsukuba near

Figure 3.7: Soyuz docked to the International Space Station. Credit: ESA

Figure 3.8: Dragon spacecraft. Credit: NASA

Figure 3.9:   Tim Peake scuba diving. Credit: ESA

Tokyo (Japan), Montreal (Canada), and at the EAC. Each of the ISS partners is in charge of providing training to all ISS astronauts on the elements which they have contributed to the ISS Program. Since it is impossible to describe all the training elements of Peake's astronaut training program, what follows is a snapshot of some of the more popular training that was featured by the media as Peake completed his transition from military pilot to fully-fledged astronaut.

Panel 3.1 Columbus laboratory

The Columbus laboratory is ESA's most significant contribution to the ISS. The 4.5-meter diameter cylindrical module is equipped with flexible research facilities, providing extensive science capabilities. The laboratory has room for ten International Standard Payload Racks (ISPRs), each the size of a telephone booth, and is capable of hosting its own autonomous and independent laboratory, complete with power and cooling systems, and video and data links back to researchers on Earth. ESA has developed a range of payload racks, each designed to maximize the amount of research, and offers European scientists full access to a weightless environment.

At launch, Columbus was fitted with five internal payload racks:

a. Biolab supports experiments on micro-organisms, cells and tissue cultures, and small plants.

Figure 3.10: Tim Peake and Samantha Cristoforetti build a shelter during survival training. Credit: ESA

b. The European Physiology Modules Facility (EPM) enables investigation into the effects of long-duration spaceflight on the human body.
c. The Fluid Science Laboratory (FSL) accommodates experiments in the behavior of weightless liquids.
d. The European Drawer Rack (EDR) is a modular experiment carrier system for a variety of scientific disciplines and provides basic resources for experiment modules housed within standardized drawers and lockers.
e. The European Transport Carrier (ETC) accommodates items for transport and stowage.

Since scientists on the ground rely on astronauts to conduct their experiments in space, the astronauts must be proficient in the operation of the equipment on Columbus, which is why training is conducted in the lab during the advanced training phase.

EVA pre-familiarization training

ESA's EAC developed the EVA Pre-Familiarization Training Program (EPFTP) to bridge the gap between scuba training (which Peake received during basic training) and NASA's EVA skills training. By completing the EPFTP, Peake developed the essential cognitive, psychomotor and behavioral skills required to prepare him to use the ISS spacesuit in NASA's Neutral Buoyancy Laboratory (NBL – Figure 3.11).

Figure 3.11:   Tim Peake training in the Neutral Buoyancy Laboratory. The NBL is 62 meters in length, 31 meters wide, and more than 12 meters deep. It contains 23.5 million liters of water in a volume big enough to submerge full-scale ISS module and payload mock-ups, as well as visiting vehicles such as JAXA's HTV, SpaceX's Dragon, and the Orbital Sciences Corporation's Cygnus. One headache with using neutral buoyancy diving as a simulation of microgravity is the amount of drag created by the water, which makes it difficult to set an object in motion and to keep it moving (this is the opposite of what is experienced in space where it takes no effort at all to set an object going). Another problem inherent in neutral buoyancy simulation is the fact that astronauts are not weightless *within* their suits. This means that as divers tilt their suits, the astronauts are pressed against whatever inside surface is facing down. After a while, this can be uncomfortable in certain orientations. Credit: NASA

The EPFTP teaches astronauts to acquire these skills in a two-week course, comprising a series of classroom courses and in-water exercises designed to challenge the astronauts while performing simulated EVAs. First, the astronauts are provided with an overview of the EMU suit, including its biomechanics and constraints in water and space. Next, instructors give the astronauts recommendations on the best strategies for moving in the suit without fighting it, and also the no-go zones on the space station. Then, in a simulated EVA session supervised by instructors, the astronauts have the opportunity to test their movement strategies underwater in EAC's Neutral Buoyancy Facility (NBF, Table 3.3). In this session, the astronauts practice different body postures, changing attitude and body orientation around the confining structures of the underwater station. After their underwater EVA indoctrination session, the astronauts are introduced to the tools they will be using outside the ISS and the interfaces to equipment that will be used in future exercises. After becoming familiar with their EVA tools, the astronauts must then be qualified in the Surface Supplied Diving System (SSDS), enabling them to communicate with the Control Room during their underwater training.

After the SSDS qualification, astronauts conduct the 'EV1 Run', akin to a real EVA. During EV1, the astronauts wear a low-fidelity mini-workstation strapped to their chests to carry EVA tools, a backpack simulating the EMU's Primary Life Support System (PLSS), a helmet, a pair of boots and EMU gloves. Guided by instructors clad in scuba-diving gear, the astronauts perform an end-to-end EVA, including airlock egress/ingress, payload transportation, use of waist tethers, and operation of ISS connectors. Throughout the session, the astronauts must comply with EVA rules, such as tethering the body at all times and using only D-rings as tethering points.

The third underwater session is the EV1+2 Run. This session is a two-member EVA with an emphasis on team situational awareness, crew communication and workload management. During this EVA, the team is free to develop their own timeline and decide how the EVA tools are shared. In addition to performing routine tasks, the astronauts must also

Table 3.3 ESA's Neutral Buoyancy Facility Characteristics

Water Tank

Length	Width	Depth	Volume	Temperature
22m	17m	10m	3,747m^3	27-29°C

Scuba and SSDS Diving Equipment

20 complete sets of scuba diving equipment (cylinders, regulators, suits, etc.)
3 SSDS sets comprising:
- Full-face mask with microphone and earphones for 2-way communication between SSDS divers and on-deck personnel.
- Buoyancy jacket including inflator, with 6 liter/300 bar reserve air tank, pressure gauge and dive computer.
- 60 m umbilical hose connected to deck air supply and for communication cables.

SSDS cart on-deck hosting umbilicals, air tanks, pressure-monitoring devices and video and audio monitoring.

EVA Tools

EVA connectors (electrical and fluid).	Limited sets of EVA and dummy tethers.
Portable Foot Restraints (PFRs) mounted on EVA worksites on Columbus mock-up.	ISS handrails mounted on airlock and Columbus mock-up.
EMU-like boots for use with PFR	EMU PLSS backpack, helmet and gloves

NBF Control Room

Video: 8 channels, including switching matrix for observation and multiple recording of underwater and deck operations.	Audio: 2 audio loops for bi-directional communications between deck personnel and between deck personnel and divers.
Additional monitor with switching matrix for deck personnel in NBF hall.	Underwater loudspeakers for unidirectional communication with all divers.

Table 3.4 Major Peake's Journey to Space

2008	Applies to ESA. Start of demanding, year-long selection process
2009	Selected to join ESA's Astronaut Corps
2010	Completes 14 months of astronaut basic training
2011	Peake and five other astronauts live in caves in Sardinia for a week.
2012	Spend 12 days living in NEEMO (underwater base in Florida). Complete training and certification for space walks
2013	Assigned a six-month mission to ISS
2015	Embarks on Principia mission

respond to unexpected equipment failures and unplanned activities, while being controlled by a Test Director. Upon completion of the EPFTP, the astronauts are given a study guide and DVD package with the course material, including videos of their EVA runs together with additional reference documentation and EVA skills demonstration videos.

While ESA's basic training phase is predominantly knowledge-based classroom training, the advanced phase includes several practical elements such as EVA, requiring the ascans to spend time in training mock-ups and simulators. On completion of the advanced training phase, an ESA Ascan is finally eligible for spaceflight assignment (see Table 3.4), which inevitably means even more training!

CAVES AND NEEMO

One such training increment involved living in a cave for six days, as part of a rather unique and elaborate analog mission to simulate some aspects of a long duration space mission. The September 2011 expedition, dubbed CAVES (Cooperative Adventure for Valuing and Exercising), took place in the Sa Grutta cave system (Figure 3.12) on the Italian island of Sardinia in the Mediterranean.

> "Even for astronauts, life in the dark, cool, humid underground environment can be a completely new situation, with interesting psychological and logistical problems. The cave environment is isolated from the outside world. There is confinement, minimal privacy, technical challenges and limited equipment and supplies for hygiene and comfort — just like in space."

Loredana Bessone, EAC astronaut trainer, Cologne, Germany

For his CAVES experience, Peake was joined by ESA astronaut Thomas Pesquet, Japanese astronaut Norishige Kanai, Russian cosmonaut Sergey Rzyhikov, and NASA astronaut Randolph Bresnik. Before heading below ground, the team spent time learning the safety protocols, cave progression techniques, and surveying and orientation skills. They were also taught cave photography so they could document their scientific work. After learning the ropes the cavers headed underground, sliding along steel cables and climbing ropes to get to base camp. Once there, they set up camp and settled into cave routine. Each morning, just as they would on ISS, they held a planning conference and reported the achievements from the previous day. And after six days of conducting experiments (mapping, monitoring

Figure 3.12: Tim Peake during survival training. This particular training took place in the underground system of Sa Grutta caves in Sardinia, Italy. The CAVES (Cooperative Adventure for Valuing and Exercising) course served as a platform to evaluate astronauts' behavior and performance while isolated. The course is run by the European Astronaut Center and both seasoned ISS crewmembers and rookies participate. During their first week living as a cave-dweller, the astronauts learn the necessary safety techniques unique to their environment, in addition to cave progression techniques and cave surveying. They are also taught cave photography so that they can document their stay and their scientific work. That takes up about 5 days. Then it's down into the bowels of the cave system for the actual expedition. The astronauts set up base camp and then they conduct day-to-day activities pretty much as they would on ISS, beginning each morning with a daily planning conference. Then, after 11 days without seeing the Sun, they head up to the surface. On exit, they experience a similar sensory overload to that of astronauts returning from ISS, although a little more muted. Credit: ESA

airflow and photography), living as a team, and coping with the language and cultural barriers (again, just like they would have to do on orbit) they returned to the surface.

"It took about five hours to come back from the cave to our campsite, requiring technical caving and a support team to help us. We really had a feeling of being far away. When we came back, everything on the surface looked strange: the blue of the sky and other colors looked painted and all the smells of nature were so strong. The real world felt all *too* real, exaggerated."

Tim Peake describing the unique CAVES space analog experience

The following year, Peake took part in another, more traditional space analog training increment, when he was assigned to the NEEMO (NASA's Extreme Environment Mission Operations) 16 mission, together with Cornell Professor Steven Squyres, habitat technician James Talacek, JAXA astronaut Kimiya Yui and NEEMO Commander Dottie Metcalf-Lindenburger. Measuring eight meters in length and submerged 20 meters below the surface of the Florida Keys, NEEMO has been an analog staple for years now (NASA leases the base – called Aquarius – from the National Oceanic & Atmospheric Administration, the NOAA). During his 10-day stay, Peake and his colleagues (Figure 3.13) worked with equipment that might be used to explore an asteroid sometime in the late 2020s. To simulate such a mission, communications was subject to a 50-second delay to see how this impacted the asteroid work.

During their stay under water, the NEEMO 16 team investigated operational and technical concepts that might be implemented in a mission to an asteroid. For example, the team investigated anchoring techniques, tested robotic systems, assessed the effects of the communications delay, and measured how mission activities could be shared among the crew. Most astronauts are in agreement that NEEMO serves as the best space exploration analog, because there is no quick emergency exit. That's because the crew-members are under pressure and become de facto saturation divers while on the base: any emergency requires decompression, so there is no immediate return to safety.

Figure 3.13: Tim Peake, Steven Squyres, James Talacek (inside), Justin Brown, Kimiya Yui (JAXA), Cdr. Dottie Metcalf Lindenburger (NASA) at the NEEMO base off the Florida coast. Credit: NASA

With CAVES and NEEMO checked off, Peake returned to astronaut duties, knowing that the next ISS expedition crew was due to be announced that autumn. And in October 2012, Peake got the call that all astronauts wait for: in a blaze of publicity, Peake was assigned to an expedition increment to the ISS in 2015– 2016. Which, of course, meant more training.

4

Britain's Astronauts

Figure 4.0:   Credit: ESA

"The International Space Station is supposed to be like a three or four-star hotel compared to the old *Mir* space station. They say it's a bit like going on a hardship camping trip where you're really having to live a bit rough. We had blackouts occasionally when we didn't have enough electrical energy in the batteries. We had an

© Springer International Publishing AG 2017
E. Seedhouse, *Tim Peake and Britain's Road To Space*, Springer Praxis Books,
DOI 10.1007/978-3-319-57907-8_4

oxygen valve that got stuck open during the launch. There's always something happening that you're there to fix. Nowadays, communications are so much better, and I think that's something all astronauts really enjoy. If you can't communicate with family and friends, you do tend to miss those human relationships much more."

Helen Sharman, 52, Britain's first astronaut, speaking at
Imperial College London.

Steve Binnie ✔ Follow
@evibenstein

How is #TimPeake Britain's first "official" astronaut? Did Helen Sharman just sneak into the spaceship when nobody was looking?

6:41 AM - 15 Dec 2015

↩ ↻ 14 ♥ 11

Emma Kennedy ✔ ✔ Follow
@EmmaKennedy

I don't give a tinker's flick who funded her. Helen Sharman is British. And was the first British astronaut. The end.

6:43 AM - 15 Dec 2015

↩ ↻ 299 ♥ 411

Until the 'Peake Era', the words 'British' and 'astronaut' were rarely seen together in the same sentence, so it's easy to forget that 24 years before Tim Peake's mission to the International Space Station (ISS), there was another British astronaut. Her name was Helen Sharman, and she *was* the first Brit in space. As with so many events that occurred in the pre-Twitter age, Sharman's story was largely forgotten amidst the fanfare surrounding Peake's mission, which is a shame, because her story is no less fascinating. Sharman launched on May 18, 1991, on board the Soyuz TM-12 mission to *Mir*, as part of a deal between Britain and the Soviet Union dubbed Project Juno (Figure 4.1). Sharman, who spent eight days on the balky outpost, not only became the first Brit in space, but also the first woman to visit the Russian space station. On her return to Earth, she was feted as a celebrity. But in 2015, when the spotlight was trained on Peake, Sharman was largely forgotten, with many media outlets mistakenly referring to Peake as Britain's first astronaut. To their credit, the UK Space Agency, modified the 'first' epithet by referring to Peake as Britain's first 'official' astronaut, alluding to his status as the first government-sponsored astronaut to fly with a space agency (ESA).

"You can't imagine how deep the color is. And the detail: you can see continents, but also the wake of a ship. And, at night, the lights of cities shine up to you.

Figure 4.1:   Mission patch for Helen Sharman's Juno flight. Public domain

There was a window where I slept, and waking up to the world right outside ...
wonderful."

Helen Sharman, reminiscing about her flight to Mir *in an interview with*
The Guardian's *Colin Drury, Monday April 18, 2016*

To Sharman, reading articles referring to Peake as the first British astronaut felt like a
snub, but to be fair, one of the reasons the intensity of her fame faded was because she led
a very private life for many years, refusing all interview requests and basically shunning
the limelight. Not surprisingly, over time, most people forgot that Britain once had an
astronaut. Over the years, there was the occasional mention of British-born astronauts who
had ventured over the pond and had acquired American citizenship in their bid to fly into
space, but these weren't true Brits in the Sharman mold.

These days, as mentioned earlier, Sharman works as operations manager of the chemistry department at Imperial College London, a place where most are unaware that one of
their colleagues was Britain's first spacefarer. After returning to Earth following her flight,
Sharman soon found the fame of her achievement to be an unwelcome downside of her
experience on orbit. She decided that she valued her privacy over the celebrity lifestyle, so
she simply disappeared from the public gaze (astronauts becoming recluses isn't as
uncommon as you might think: Neil Armstrong became a loner, Buzz Aldrin turned to
alcohol and Alan Bean turned his back on science to become an artist). Then, in 2013,
following the announcement of Peake's mission to the ISS, Sharman discovered she was
being written out of history by the UK Space Agency, who listed Peake as Britain's first
astronaut. She promptly reminded the agency that she had flown as part of the Soviet
Union's space program and the agency soon amended Peake's title to include the word
'official'. She didn't hold a grudge against Peake though; she handed the ESA astronaut a
copy of Yuri Gagarin's '*Road to the Stars*' before he blasted off, and encouraged him to
look out of the window as much as he could during his stay on orbit.

HELEN SHARMAN

Sharman was working as a PhD student and research chemist at Birkbeck University in London when she heard the '*Astronaut wanted, no experience necessary*' strapline for Project Juno. The project was partly conceived to give a shot in the arm to 'London – Moscow' relations by sending a British citizen to *Mir*. Some 13,000 astronaut wannabes applied and for those who made it to the final stages, myriad medical exams, psychological tests and aptitude tests awaited. Ultimately, just two – Sharman and Tim Mace – were selected for 18 months of cosmonaut training at Star City, with Sharman eventually given the nod as prime crew.

> "I can only surmise why me. I was physically fit, good in a team and not too excitable, which was important. You can't have people losing it in space. I think it was just my normality."
>
> *Helen Sharman, explaining why she was selected*
> *for her flight to Mir, in an interview with*
> *The Guardian's Colin Drury, Monday April 18, 2016*

For British space enthusiasts, Project Juno was a reason to celebrate, because for the best part of a decade, Margaret Thatcher's Conservative government had systematically gutted any hope of a manned space program. Before Juno, the only bright light in what was a dark age for the non-existent British human space program was the plan to fly Squadron Leader Nigel Wood as a Payload Specialist on board the Shuttle, to observe the deployment of a British military communications satellite. Wood was selected in May 1985 and preparations seemed to be on target for the first British astronaut to launch the following year, until the tragic loss of *Challenger* and her crew in January 1986 put an end to any plans for Wood (and many others) to fly.

Following Sharman's selection, the British public was keen to learn more about their first astronaut. Sharman, they soon discovered, had never dreamt of becoming an astronaut. Born in a suburb of Sheffield in 1963, she led a remarkably normal life, gaining a degree in chemistry from the University of Sheffield in 1984 before working as an engineer for the General Electric Company (GEC) in London. It was GEC who gave her the chance of pursuing her PhD, which led Sharman to enroll at Birkbeck College. That move eventually led to her working for Mars Confectionary, developing ice-cream products. It was while driving home one evening in June 1989 that she heard the radio call for astronaut candidates. After the inevitable filtering of the candidates that is par for the course in any astronaut selection, Sharman found herself among the final 32 selectees at Brunel University, where she met Royal Army Air Corps helicopter pilot, Major Tim Mace[1], the candidate that Sharman reckoned was the standout choice. Further rounds of screening reduced the 32 candidates down to 22 and then 16, with Sharman (and Mace) still in the hunt. The 16 was then reduced to just 10 and, after the final round of medical tests had been taken care of, the group became just six. Eventually, Mace (Figure 4.2) and

---

[1] For those who like trivia, Mace went on to work as South African President Nelson Mandela's helicopter pilot.

Figure 4.2: Tim Mace. After coming within a whisker of flying in space to *Mir*, Mace returned to his life's passion of sky-diving. He contested more than 150 competitions, including several World Championships in Formation Skydiving, Freefall Style, Accuracy Landing and Canopy Piloting. He was the joint holder of the 400-person FAI Formation Skydiving Large Formation World Record. In 2012, he was awarded the FAI Gold Parachuting Medal. Mace died from cancer in 2014. Credit: Russian Space Agency

Sharman were the two candidates chosen to take the flight to Star City to commence training for the Juno flight.

While Mace and Sharman were training, Project Juno hit the first, and probably the most significant, of many stumbling blocks that threatened to derail the mission. Project Juno was a commercial venture with a price tag of $12 million, a fee that the Soviets had imposed. But the actual funding raised from sponsorship was far short of that figure, despite British company Antequera Ltd pooling together a consortium of sponsors that included British Aerospace, Memorex and Interflora. When it was revealed that Antequera had raised just $1.7 million, the British government was approached to provide the funding shortfall. But the government refused to support the mission. This was December 1990, by which time Mace and Sharman had completed most of their training. Fortunately, work was in progress behind the scenes to salvage the mission, which was eventually saved by the Moscow Norodny Bank[2].

From an international perspective, Britain couldn't have been more humiliated. One of the world's richest and most technologically advanced nations had just had its first foray into the manned spaceflight arena, and had failed miserably. For Britain, $12 million[3] was a financial drop in the ocean, but the government simply wasn't interested in backing the mission, apparently preferring instead to deal with the shame, ignominy and disgrace of that decision rather than supporting international relations. It was a decision that further underlined the British government's intransigence and underscored the country's standing as the only major economic power without any interest in human spaceflight. Officially, the British government's stance on manned spaceflight was that such an endeavor had to demonstrate user benefits and not just be a national status symbol. The government also pointed out that the country had preferred to invest in robotic missions, ignoring the fact that robots do little to fire the public imagination.

While the British government buried its head deeper in the sand, the decision about who would be the first Briton in space was looming. That choice was made on February 19, 1991, when Sharman was informed that she would be the prime candidate. The actual mission was just three months away. On May 18, 1991, Sharman and her Russian colleagues boarded the Soyuz TM-12, whose nose cone had been painted with the Hammer and Sickle and the Union Jack. Crammed into the claustrophobic confines of the Soyuz vehicle with Sharman were Anatoli Artsebarski, on what would be his only space flight, and Sergei Krikalev (Figure 4.3), who was embarking on the second of his six missions to Earth orbit. Liftoff occurred at 03:50 Moscow time and two days later, on May 20, the Soyuz was in position for the final stages of its approach to *Mir*. What followed should have been an automated procedure, but the balky KURS rendezvous system developed a glitch, forcing Artsebarski to fly the approach manually. After docking, the resident crew of Viktor Afanasyev and Musa Manarov welcomed the incoming crew with the traditional bread-and-salt greeting, and Sharman went to work on her depleted experiment program. That program should have featured British experiments, but thanks to the miserable failure of British commercial participation, her program now consisted almost exclusively of Soviet science. For eight days, Sharman wore electrodes to measure her heart rate, measured her reaction time using light patterns, and took blood samples to assess her adaptation to microgravity. In addition to the life sciences experiments, Britain's first spacefarer also grew potato roots to study the effect of radiation on genetic mutations. In all, Sharman tended to 17 biotechnological and medical experiments. All of which were Soviet. In fact, the only British component of the experiments was the person doing the testing.

Sharman returned to Earth on May 26, together with Afanasyev and Manarov on board the Soyuz TM-11 vehicle. On landing, Manarov became the record-holder for longest time in space, having amassed a cumulative 541 days. He would go on to a political career as a member of the Russian Duma. Sharman, meanwhile, had her sights on returning to space and applied to ESA's astronaut selections in 1992 and 1998 (in 1992 she was

---

[2] And by the intervention of Soviet Premier Mikhail Gorbachev, if the rumors are to be believed.

[3] In 1992, Germany, struggling with the horrendous cost of reunification, was still able to find more than $20 million to fly one of its citizens as a cosmonaut.

Figure 4.3: The Soyuz TM-12 crew: Sharman, Artsebarski and Krikalev. Credit: Russian Space Agency

one of three British candidates and in 1998 she was on the shortlist of 25 applicants) without success. Following her flight, she became one of Britain's leading ambassadors for science and was awarded the OBE in 1992. Until Peake's selection, Britain's venture into manned space flight had stalled for more than two decades, although a few Britons did follow in Sharman's footsteps, albeit without the sanction of the government. We'll introduce them in this chapter.

MICHAEL FOALE

"Britain's exploration history is huge. It stops somewhere in the middle of the last century and I would like to see it pick up again; and I think Tim represents that."

Dr. Michael Foale, voicing his opinion of the UK government's lack of support for a manned spaceflight program in an interview with the BBC in 2009.

Figure 4.4: Michael Foale. Credit: NASA

While sometimes critical of the British government's attitude to manned spaceflight, Dr. Foale was delighted when ESA selected Peake to the astronaut corps.

By the time Colin Michael Foale hung up his spacesuit in 2013, only two NASA astronauts had spent more time on orbit than him. During what became one of the most illustrious of all astronaut careers, Dr. Foale accumulated 374 days in space, made six trips to low Earth orbit (LEO), performed four spacewalks, *and* helped save the stricken *Mir* after a Progress cargo vehicle collided with the aging Russian space station. Like all 'British' astronauts who flew into LEO after the Sharman era, Foale was only able to follow his chosen career path thanks to a change in passport status. In Foale's case, his ticket to astronaut selection came about thanks to his mother being an American citizen. Born in Louth, Lincolnshire, in 1957, Foale's path to astronaut selection began with a first-class honors degree in natural sciences, followed by a doctorate in astrophysics from Queen's College, Cambridge, and a pilot's license courtesy of time spent in the Air Training Corps. After securing a job with NASA, Foale worked at Johnson Space Center's (JSC) Mission Operations Directorate in 1983, where he had the opportunity to work as a payload officer on Shuttle missions STS-51G, 51-I, 61-B and 61-C. While at JSC, Foale applied to the

Table 4.1 Table STS-45

Mission	
Mission type:	Research
Mission duration:	8 days, 22 hours, 9 minutes, 28 seconds
Distance travelled:	5,211,340 kilometers
Orbits completed:	143
Spacecraft	
Spacecraft:	Shuttle *Atlantis*
Landing mass:	93,009 kilograms
Payload mass:	9,947 kilograms
Crew	
Crew size:	7
Members:	Charles F. Bolden, Jr.
	Brian Duffy
	Kathryn D. Sullivan
	David C. Leestma
	C. Michael Foale
	Dirk Frimout
	Byron K. Lichtenberg
Launch	
Launch date:	March 24, 1992, 13:13:39 UTC
Launch site:	Kennedy LC-39A
Landing	
Landing date:	April 2, 1992, 11:23 UTC
Landing site:	Kennedy SLF Runway 33
Orbital parameters	
Perigee:	282 kilometers
Apogee:	294 kilometers
Inclination:	57.0 degrees
Period:	90.3 min

astronaut corps twice and was twice rejected, before succeeding on his third attempt in 1987. His first flight was as a Mission Specialist aboard *Atlantis* on STS-45 (Table 4.1), which launched on March 24, 1992.

STS-45, which carried the Atmospheric Laboratory for Applications and Science (ATLAS-1), was a science mission that featured experiments in solar radiation, plasma physics, atmospheric chemistry and ultraviolet astronomy. These experiments included:

- *The Imaging Spectrometric Observatory (ISO)*, which measured the reactions and energy transfer processes that occur in the atmosphere.
- *The Atmospheric Trace Molecule Spectroscopy (ATMOS)* experiment that measured trace molecules in the middle atmosphere.
- *The Atmospheric Lyman-Alpha Emissions (ALEA)* experiment that observed ultraviolet light (Lyman-Alpha) in which hydrogen and deuterium radiate at different wavelengths.

Figure 4.5: STS-45 Mission Patch. Credit: NASA

- *The Millimeter-Wave Atmospheric Sounder (MAS)* experiment, that helped scientists better understand the distribution of water vapor and chlorine monoxide in the upper atmosphere.
- *The Shuttle Solar Backscatter Ultraviolet (SSBUV)* experiment, which measured ozone levels and compared this data with measurements taken on board NOAA satellites.
- *The Atmospheric Emissions Photometric Imaging (AEPI)* and *Space Experiments with Particle Accelerators (SEPAC)* experiments, that collected data on the charged particle and plasma environment in the Earth's atmosphere.

In addition to the extensive suite of radiation and solar physics experiments, STS-45 also carried life sciences experiments that included:

- *The Radiation Monitoring Experiment (RME)* that measured the crew's exposure to ionizing radiation.
- *The Visual Function Tester II Experiment (VFT-II)* that assessed contrast ratio threshold in astronaut vision.

To keep up with the busy science schedule, the crew was divided into Red (Leestma, Foale, Lichtenberg) and Blue (Bolden, Duffy, Sullivan, Frimout) teams. After more than eight days in space, *Atlantis* landed at Kennedy Space Center on April 2, 1992.

Foale's next ride into space came just over a year later, on board *Discovery* for STS-56. Like STS-45, STS-56 was a science-heavy mission. *Discovery's* primary payload was Atlas-2, the successor to the payload carried on Foale's previous mission, and there was also a full suite of support payloads stashed in the U-shaped Spacelab pallet stowed in the Shuttle's cargo bay. Provided by ESA, Spacelab functioned almost autonomously, thanks to a container (the 'igloo') that housed the payload's power supply, data-handling system, and temperature control system. In common with Foale's first flight, much of the crew's time on orbit was spent tending to the experiments, most of which were follow-ups to those conducted during STS-45.

Figure 4.6:   STS-56 Mission Patch. Credit: NASA

Among the work that wasn't carried out in Spacelab was SPARTAN – the Shuttle Point Autonomous Research Tool for Astronomy – which was a free-flying payload designed to study the velocity of solar wind and observe the Sun's corona. Another non-Spacelab experiment was HERCULES (Hand-held, Earth-oriented, Real-time, Cooperative, User-friendly, Location-targeting and Environmental System), which operated a little like a GPS system by determining real-time latitude and longitude to an accuracy of within two miles. As on STS-45, STS-56 (Figure 4.6) also flew the RME experiment, this time together with a Tissue Equivalent Proportional Counter (TEPC) that served as a micro-dosimeter for measuring accurate doses of radiation. When not tending to their myriad science tasks, the STS-56 crew spent time talking to schools and ham operators around the world courtesy of the Shuttle Amateur Radio Experiment (SAREX – Foale's call sign was KB5UAC)[4].

In addition to the numerous solar-based experiments, STS-56 (Table 4.2) supported more than 30 life sciences investigations, ranging from cell biology to crystal growth and from biotechnology to bone development. One such experiment was the Space Tissue Loss (STL) experiment that was designed to help researchers better understand the mechanisms governing bone loss in microgravity. A related life sciences experiment was the Physiological and Anatomical Rodent Experiment (PARE) that flew in the Shuttle's mid-deck. As the acronym suggests, the PARE investigation made use of rats for the purpose of studying bone loss – in this case looking at how microgravity affected the production of osteoblasts.

Foale's third mission took place less than two years after his return from STS-56, when he once again boarded *Discovery* (on its 20th mission) on February 3, 1995, for STS-63 (Figure 4.7). The first Shuttle mission of 1995 was notable for a number of reasons. One was that it was the first flight of a female Shuttle pilot (Eileen Collins), and another was that it included the first approach and fly-around of *Mir* by a Shuttle (see sidebar).

---

[4] STS-56 marked the first time that radio contact between *Mir* and the Shuttle was achieved using amateur radio equipment.

Table 4.2 STS-56

Mission	
Mission type:	Research
Mission duration:	9 days, 6 hours, 8 minutes, 24 seconds
Distance travelled:	6,202,407 kilometers
Orbits completed:	148
Spacecraft	
Spacecraft:	Shuttle *Discovery*
Landing mass:	93,683 kilograms
Payload mass:	7,026 kilograms
Crew	
Crew size:	5
Members:	Kenneth D. Cameron
	Stephen S. Oswald
	C. Michael Foale
	Kenneth D. Cockrell
	Ellen Ochoa
Launch	
Launch date:	April 8, 1993, 05:29:00 UTC
Launch site:	Kennedy LC-39B
Landing	
Landing date:	April 17, 1993, 11:37:19 UTC
Landing site:	Kennedy SLF Runway 33
Orbital parameters	
Perigee:	291 kilometers
Apogee:	299 kilometers
Inclination:	57.0 degrees
Period:	90.4 min

Figure 4.7: STS-63 Mission Patch. Credit: NASA

Table 4.3 STS-63

Mission	
Mission type:	Research and 'near *Mir*' rendezvous
Mission duration:	8 days, 6 hours, 28 minutes, 15 seconds
Distance travelled:	4,816,454 kilometers
Orbits completed:	129
Spacecraft	
Spacecraft:	Shuttle *Discovery*
Landing mass:	95,832 kilograms
Payload mass:	8,641 kilograms
Crew	
Crew size:	6
Members:	James D. Wetherbee
	Eileen M. Collins
	Bernard A. Harris Jr.
	C. Michael Foale
	Janice E. Voss
	Vladimir G. Titov (Russia)
EVAs	
Total EVAs:	1
Total duration:	4 hours, 39 minutes
Launch	
Launch date:	February 3, 1995, 05:22:04 UTC
Launch site:	Kennedy LC-39B
Landing	
Landing date:	February 11, 1995, 11:50:19 UTC
Landing site:	Kennedy SLF Runway 15
Orbital parameters	
Perigee:	291 kilometers
Apogee:	299 kilometers
Inclination:	51.6 degrees

The maneuver served as a test-bed prior to the commencement of regular Shuttle-*Mir* dockings later that year. In addition to these significant milestones, the STS-63 crew (Table 4.3) was also tasked with conducting Spacehab experiments, two spacewalks, and the deployment of the SPARTAN spacecraft.

Shuttle - Mir Rendezvous Sequence
The Shuttle's approach to *Mir* began nine hours into STS-63 when the reaction control system (RCS) fired, adjusting *Discovery's* closing rate to the Russian station. Over the next few days, further firings of the RCS brought the Orbiter to just 15 kilometers behind *Mir*. This marked the final stage of the rendezvous. With *Discovery's* rendezvous radar system providing closing rate data to the crew, the Shuttle closed the final 15 kilometers until it reached a point 609 meters below *Mir*, at which point

Commander James Wetherbee took manual control and guided the Shuttle to the intersect vector of *Mir* in a maneuver known as the V-bar (a line that follows *Mir's* direction of travel). At 121 meters in front of *Mir*, Wetherbee brought *Discovery* to a stop and waited for the "go" from Russian Flight Control Teams. Following approval, Wetherbee gently pulsed the RCS, inching the Shuttle to just 9 meters from *Mir* and aligning the Orbiter's docking module with *Mir's* docking port. The Shuttle then backed away and began a fly-around.

The day after the rendezvous, the crew deployed the SPARTAN free-flying retrievable platform. Using the Shuttle's robotic arm, Russian cosmonaut Vladimir Titov lifted SPARTAN from the payload bay and sent the spacecraft on its 40-hour mission to observe celestial targets, after which the spacecraft was caught by the arm and brought back into the payload bay. Following the SPARTAN deployment and retrieval, it was time for the mission's sole spacewalk, which was performed by Bernard Harris and Foale (Figure 4.8). The spacewalk, which lasted 4 hours and 39 minutes, was conducted to gain experience of manipulating large masses, one of which was the SPARTAN spacecraft. The pair successfully completed this task, with Foale grabbing the vehicle before handing it over to Harris, who performed various rotation and translation maneuvers. These maneuvers were then repeated by Foale while he was attached to the robotic arm.

Figure 4.8:   Michael Foale and Bernard Harris preparing for their EVA on STS-63. Credit: NASA

With the flight of SPARTAN concluded and the completion of the spacewalk, attention turned to the Spacehab module and its myriad experiments (11 biotechnology, three advanced materials, and four technology demonstrations). Many experiments, such as the Astroculture (ASC) plant growth experiment, had made an appearance on orbit before. ASC was flying for the fourth time to validate the performance of plant growth to see if certain plants might be used in a future life support system. In tandem with ASC was Chromex-6, a plant growth study that focused on the role of enzymes. While the plant growth experiments attracted some media coverage, it was a different kind of investigation that captured most of the attention among those who followed Shuttle flights. The Shuttle Glow Experiment (GLO-2) was being flown to investigate the strange glow phenomenon that had been observed by astronauts on previous missions. It had been suggested that the glow was caused by atmospheric gases striking the Shuttle (it was later determined that the glow was caused by atomic oxygen reacting with nitric oxide on the surface of the Shuttle). Another popular experiment was the Orbital Debris Radar Calibration System-II (ODERACS-II), an investigation comprising three spheres and three dipoles that were released from the payload bay along the Shuttle's velocity vector, for the purpose of being observed and tracked using ground-based radar.

The Three Bears folktale

After returning from what was often dubbed the "near *Mir*" mission, Foale found himself in Star City working on spacewalk challenges. It was October 1995, and he had not yet been assigned his next mission. That situation changed following unique circumstances that became known within NASA as the 'Three Bears' tale. At the time, NASA was working on the NASA-5 mission to *Mir* which was expected to be flown by either Scott Parazynski or Wendy Lawrence. The problem was that neither astronaut fitted the size restriction, and therefore the safety parameters, of the new Soyuz. Parazynski was too tall and Lawrence was too short. But Foale was 'just right', went the tale, which is why he found himself moving his family to Star City in order to train for a *Mir* mission.

STS-84 (Table 4.4) was the sixth[5] of what would eventually be a series of nine Shuttle/*Mir* docking missions. As part of this program, American astronauts launched and landed on a Shuttle but served as crewmembers on *Mir*, while their Russian counterparts used the Soyuz as their mode of transport. Foale's mission marked the fourth transfer of an American crewmember to the Russian orbiting outpost, with Foale exchanging places with Jerry Linenger who had flown up to *Mir* on STS-81.

> "The willingness to undergo something very different and foreign. It was that trepidation — but interest nonetheless — to get through it. To go and do this strange thing. I think it comes out of a person, based on his background, culture, and family. I'm not sure it's something we could train into a person."

> *Michael Foale, musing on his decision to serve*
> *as a crewmember on Mir.*

---

[5] The previous five missions were STS-71, STS-74, STS-76, STS-79 and STS-81.

Table 4.4 STS-84

Mission	
Mission type:	Shuttle-*Mir*
Mission duration:	9 days, 5 hours, 20 minutes, 47 seconds
Distance travelled:	6,000,000 kilometers
Orbits completed:	144
Spacecraft	
Spacecraft:	Shuttle *Atlantis*
Landing mass:	100,285 kilograms
Payload mass:	2,500 kilograms
Crew	
Crew size:	7 (including exchange)
Members:	Charles J. Precourt
	Eileen M. Collins
	Jean-Francois Clervoy (ESA)
	Carols I. Noriega
	Edward T. Lu
	Yelena V. Kondakova (Russia)
	C. Michael Foale (up only)
	Jerry Linenger (down only)
Launch	
Launch date:	May 15, 1997, 09:07:48 UTC
Launch site:	Kennedy LC-39B
Landing	
Landing date:	May 24, 1997, 13:27:44 UTC
Landing site:	Kennedy SLF Runway 33

Figure 4.9: STS-84 Mission Patch. Credit: NASA

A crisis on *Mir*

By 1997, the Shuttle-*Mir* program had matured and the complex had grown since Foale had seen it last. The space station now included both a science module and docking module. As *Atlantis* approached, Foale was struck by the increased size of the station and also by the good condition the outpost appeared to be in. This impression was boosted when he entered the station and found the place to be very welcoming, despite the tangle of cables. It would prove to be a short-lived illusion. Greeting Foale was NASA-4 astronaut Linenger, who explained that the reason the station looked so presentable was because every waking moment over the past few weeks had been spent cleaning it! After showing Foale how to activate the fire extinguisher (Linenger had been on board *Mir* when a fire occurred) and how to don a respirator, Linenger advised him that once the Shuttle departed, the illusion would be over and life aboard *Mir* would return to what it had been. Which was anything but rosy according to the departing NASA astronaut.

From the outset, and in contrast to the stand-offish Linenger, the affable Foale integrated well with the resident crew. For the first few weeks of the flight, the mission went swimmingly, although Foale needed some time to adapt to *Mir's* myriad quirks. One of these was the noise created by the Spektr module, which was where Foale slept. Foale likened sleeping in Spektr to sleeping in a garage, due to the sounds of the panels' drive motors as they rotated the solar arrays attached to the module. But apart from the sleep deprivation, life on board *Mir* was going well for Foale.

Towards the end of June, the crew prepared for the arrival of a Progress cargo vehicle. Before it arrived, however, the crew had to perform some tests, including operation of a new remotely-controlled docking system. Mission Control decided that the best way to test the system was to re-dock a Progress that had recently been un-docked from *Mir*. It was a procedure that concerned Commander Vasily Tsibliyev, because a similar procedure using a Soyuz in January 1994 had resulted in the vehicle bumping into *Mir*. Another remote docking had been attempted during Linenger's stay on the station, which resulted in loss of control over the vehicle and only narrowly avoided ending in a collision. So, on June 25, with some trepidation, Tsibliyev took remote control of the cargo vehicle and commanded the Progress to head towards the station. He followed the path of the Progress on a video screen, but that didn't allow *Mir's* commander to judge speed or distance very well, and by the time he realized the Progress was moving too fast, there was no time to avoid the inevitable collision. Realizing the impending danger, Tsibliyev ordered Foale to get inside the Soyuz and prepare for evacuation.

But it was too late.

"I watched this black body covered in spots sliding past below me. I looked closer, and at that point there was a great thump and the whole station shook."

Flight Engineer Aleksandr Lazutkin, describing the oncoming Progress

The Progress collided with the Spektr module, breaching the hull. Inside *Mir,* the crew felt the impact, followed by the hissing of air escaping and the sound of the decompression alarm. As the air rushed out of the station, *Mir* went into a spin, and the crew realized they were in a desperate and potentially fatal situation. The crew didn't know where the puncture was, but they knew it was the Spektr module that had been breached and that meant they had to seal the module off from the rest of *Mir*. That would be no easy task, because the hatch between *Mir* and Spektr was blocked by a rats-nest of tubes and cables. Simply pulling at the cables and hoses would not solve the problem, because many of them needed to be disconnected. The crew eventually cleared the hatch and succeeded in sealing it with a cover. But they weren't out of the woods yet. Some of the disconnected cables had supplied electrical power generated by Spektr's solar arrays to the rest of *Mir* and now, without that power, the primary computer shut down, plunging the station into darkness. For the crew, it was an eerie feeling, not just because of the darkness, but because the once noisy station had fallen completely silent.

> "It crossed my mind: 'You know, I've been here six weeks and I think we're going to be going home right now.' I was actually kind of sad. I thought, 'Well, you know, that's a shame. I won't finish this whole thing. I had set out to be here four and a half months, and now it's going to get cut short. This is a real emergency.' And we had all the danger of getting out of there, but it was crossing my mind, 'This is a shame. I've only been here six and a half weeks. What a shame I'm not getting to do the whole thing. Then it occurred to me, 'Well, you know, you'll get to see your kids and Rhonda [his wife] sooner.' And I thought, 'Oh, but we're going to be landing in Kazakhstan. That's going to be a delay.' The thoughts that went through my mind, it was exactly like that. I thought, 'You'd better focus on getting this sealed off here.' That's what went through my mind."

> *Michael Foale, reminiscing about his time on Mir*
> *for NASA's Oral History Project.*

The first order of business for the crew was to determine the spin rate of the station, but this proved difficult without the computer. Fortunately, astronauts possess well above average initiative, as Foale demonstrated. He estimated the spin rate – by applying basic practical naval navigation and a scientist's knowledge of physics – by holding his thumb up to the field of stars. He then radioed down his estimate to Moscow's Control Center and the ground controllers fired the station's engines to slow the spin. It wasn't a perfect solution, but it was a good start. Over the next few hours, Foale continued his star observations and passed on instructions to Tsibliyev, who was in the Soyuz. On receiving Foale's information, Tsibliyev applied pulses of the spacecraft's reaction control system to establish a stable rotation and point the solar arrays towards the Sun. With that job done, the crew hunkered down for some much-needed shuteye.

The next day, the crew took stock of their situation. With Spektr out of commission, Foale had not only lost his bedroom, but also all of his personal items and experiments. That problem was solved by a care package put together by NASA, which was sent up on the next Progress vehicle that arrived and docked – without incident this time – on July 7. The Progress also ferried up a modified hatch cover, designed to allow all the disconnected cables to be reattached. That task was to be conducted by a spacewalk *inside* Spektr. The spacewalk was due to be performed by Tsibliyev and Lazutkin, but the cursed crew was dealt another blow a week after the arrival of the Progress when Tsibliyev recorded an irregular heartbeat while exercising. This disqualified him from spacewalking duties, and meant that Foale was now in the frame to take Tsibliyev's place. So Foale began training for the upcoming spacewalk, and for a while life appeared to be returning to normal on board the jinxed station. That bubble burst on July 17, when Lazutkin accidentally disconnected a vital power cable. For a cosmonaut wracked by exhaustion, mistakenly disconnecting the wrong cable that was one of hundreds was forgivable, but Lazutkin was inconsolable:

> "Oh, my feelings! Shooting yourself would be easier! It was terrible. The station emergency alarm went off. I realized instantly that I'd made a mistake."
>
> *Lazutkin, explaining his mistake on board Mir in PBS's*
> *Terror in Space (air date October 27, 1998).*

The fallout of the cable disconnection was another power outage and failure of the station's computer, with the result that *Mir* was once again sent tumbling across LEO. With the station in a state of crisis, it was decided to delay the spacewalk until the *Mir*-24 crew arrived in August. Meanwhile, Foale attempted to salvage what he could of his mission. Not surprisingly, the Progress collision had put a big dent in his investigations, but there were some salvageable experiments, such as those in the greenhouse that had miraculously survived intact.

Following the arrival of the *Mir*-24 crew of Anatoly Solovyev and Pavel Vinogradov, the two crews performed the hand-over and prepared for Tsibliyev and Lazutkin's departure. The first major task for the new crew was the spacewalk performed by Solovyev and Vinogradov to restore power to the Spektr. This was to be followed by a second spacewalk, conducted by Foale and Vinogradov, to inspect the stricken module for damage. During the first spacewalk, Solovyev and Vinogradov's inspection of Spektr revealed the location and size of the puncture, but apart from the breach, the module was in good shape. After reconnecting power from the solar arrays, the two spacewalkers retrieved Foale's laptops and headed back into the station. A few days later, it was Foale and Solovyev's turn to spend time outside *Mir*. During their six-hour excursion, the two spacewalkers observed plenty of damage to Spektr but no hull breach.

With power now reconnected, *Mir* gradually returned to its pre-crisis state, and Foale's mission approached its conclusion. Meanwhile on the ground, where *Atlantis* was being readied to collect Britain's first spacewalker, the fallout from the crises that Foale and his colleagues had endured inevitably caused some to question whether the next NASA astronaut due on the station (David Wolf) should replace Foale at all, given all the trials and tribulations. Foale responded to the hullabaloo with an upbeat statement:

> "I'd like to summarize really why I think Dave Wolf should stay onboard space station *Mir* when I leave. Really, I think it comes down to the fact that, even though this flight has been one of the hardest things I have ever attempted in my life, I have to remember what John F. Kennedy said when I was about four years old. Forgive me if I get it wrong. He said, 'We do not attempt things because they are easy, but because they are hard, and in that way we achieve greatness.' I believe out of this cooperation of America with Russia, which is not always easy, we are achieving some extremely great things. And, for these reasons, I think I've really valued my

Figure 4.10:   STS-103 Mission Patch. Credit: NASA

time onboard space station *Mir*. I will always remember the last three or four months with great, great alacrity and nostalgia, I'm sure. I really count all that we are doing together, America and Russia, to be extremely valuable to future cooperation on the Earth in the future."

Foale was picked up by *Atlantis*, flying her 20th mission, on September 27, 1997. The hatches between the Shuttle and *Mir* were closed on October 3, and soon after his return to Earth, Foale found himself searching for a new challenge. That came in the form of the STS-103 mission (Figure 4.10), a Hubble Space Telescope (HST) repair and servicing mission (to replace six gyroscopes) slated for the end of the millennium.

Servicing Hubble

Foale's HST servicing mission became one of the most troublesome missions on NASA's manifest, initially at least. After having been delayed repeatedly by technical problems, the launch was scrubbed three times, by further technical problems and weather issues. But *Discovery* finally launched on December 19, 1999, and after a 30-orbit chase, Mission Specialist Jean-Francois Clervoy grappled the Hubble using the Canadarm and maneuvered it into the Shuttle's payload bay. The crew began preparing for the first of what would be three spacewalks. First out of the airlock were Mission Specialists John Grunsfeld and Steven Smith, who spent eight hours and 15 minutes outside (the second longest EVA during the Shuttle program up to that point) installing voltage/temperature improvement kits between HST's solar panels. The next pairing was Foale (Figure 4.11) and Swiss astronaut Claude Nicollier, who installed a new computer and guidance sensor during another marathon extravehicular excursion that lasted eight hours 10 minutes. The third and final spacewalk was performed by Grunsfeld and Smith again, adding another eight hours and eight minutes to their EVA scorecard as they installed a transmitter to enable better data transmission. With STS-103 under his belt (Table 4.5), Foale had notched up his fifth spaceflight, but he wasn't ready to retire yet. One item missing from his already stellar resume was a command position. That opportunity presented itself in the form of Expedition 8.

Figure 4.11:   Michael Foale, STS-103. Credit: NASA

Table 4.5 STS-103

Mission	
Mission type:	Hubble Service
Mission duration:	7 days, 23 hours, 11 minutes, 34 seconds
Distance travelled:	5,230,000 kilometers
Orbits completed:	119
Spacecraft	
Spacecraft:	Shuttle *Discovery*
Launch mass:	112,493 kilograms
Landing mass:	95,768 kilograms
Crew	
Crew size:	7

(continued)

Table 4.5 (continued)

Members:	Curtis L. Brown Jr.
	Scott J. Kelly
	John M. Grunsfeld
	Jean-Francois Clervoy (ESA)
	C. Michael Foale
	Steven L. Smith
	Claude Nicollier (ESA)
EVAs	
Total EVAs:	3
Total duration:	24 hours, 33 minutes
Launch	
Launch date:	December 20, 1999, 00:50:00 UTC
Launch site:	Kennedy LC-39A
Landing	
Landing date:	December 28, 1999, 00:01:34 UTC
Landing site:	Kennedy SLF Runway 33
Orbital parameters	
Perigee:	563 kilometers
Apogee:	609 kilometers
Inclination:	28.45 degrees
Period:	96.4 min

ISS commander

Expedition 8 (Figure 4.12) afforded Foale his first visit to the relatively new ISS, which had been born on November 20, 1998 with the launch of the Zarya base module. Launched on October 18, 2003, Soyuz TMA-3 carried Expedition 8 Commander Foale, Expedition 8 Flight Engineer Alexander Kaleri, and ESA astronaut Pedro Duque, who was flying a short-duration flight (mission name *Cervantes*) born of a commercial agreement between ESA and Rosaviakosmos, the Russian space agency. Following the standard two-day ISS approach, the Soyuz docked with the station on October 20, after which Foale and Kaleri officially replaced the Expedition 7 crew of Commander Yuri Malenchenko and NASA ISS Science Officer Ed Lu, who had been aboard the orbiting outpost since April 2003. From Expedition 7, the ISS resident teams had been reduced to 'caretaker' crews of just two, because the Shuttle fleet had been grounded following the loss of *Columbia*, leaving the smaller Soyuz as the only means of transport. After settling into their new home, Foale and Kaleri familiarized themselves with the lengthy list of science investigations scheduled to be performed during their tour of duty. One experiment that attracted a lot of attention was Matroshka (see sidebar, Figure 4.13), a 'human phantom' designed to measure exposure to radiation.

Figure 4.12:   Expedition 8 Mission Patch. Credit: NASA

Phantoms in Space

One of the most serious mission-killers facing a manned voyage to Mars is long-term exposure to deep-space radiation, and attempts to address the problem have included some rather unusual experiments. The strangest of these is undoubtedly Matroshka, various versions of which have been flown in space since August 1989. Matroshka 1.0, aka the Phantom Head, made its debut in the space life sciences arena on board Shuttle flight STS-28 and later flew on STS-36. Constructed using a real human skull, the Phantom Head was fitted with 400 thermoluminescent detector chips (TLDs) to measure radiation doses, specifically those generated by galactic cosmic rays (GCRs). During its flights, the Phantom Head was tucked away in a middeck locker shrouded by a white Nomex bag, which just added to its already eerie appearance (Shuttle crews dubbed the head 'Satan'). After each flight, the head was returned to the Phantom Lab, where it was disassembled so that data from the dosimeters could be downloaded. The Phantom Head was a success, which spurred investigators to create a larger phantom to represent a head and torso. Officially known as the Phantom Torso, the new space phantom was dubbed Fred by astronaut crews. Fred made his first appearance in LEO on board *Discovery* during the STS-91 mission and returned to space with the crew of STS-100, where he was placed inside one of the racks in the new Destiny module on ISS. Sliced into 34 segments (which inspired the moniker 'Matroshka', derived from the Russian nested dolls known as *matryoshka*), each of which was fitted with passive and active radiation sensors, Fred provided researchers with a wealth of data about radiation levels inside the ISS before returning to Earth with the STS-105 crew. Logically, the next step was to characterize the radiation environment *outside* the ISS, which is how Mr. Rando came into being. Constructed of natural bone, polyurethane to simulate organs, and carbon fiber to simulate an astronaut's spacesuit, Mr. Rando (the trademark name of the torso) was installed on the exterior of the Zvezda module during Expedition 8's only spacewalk.

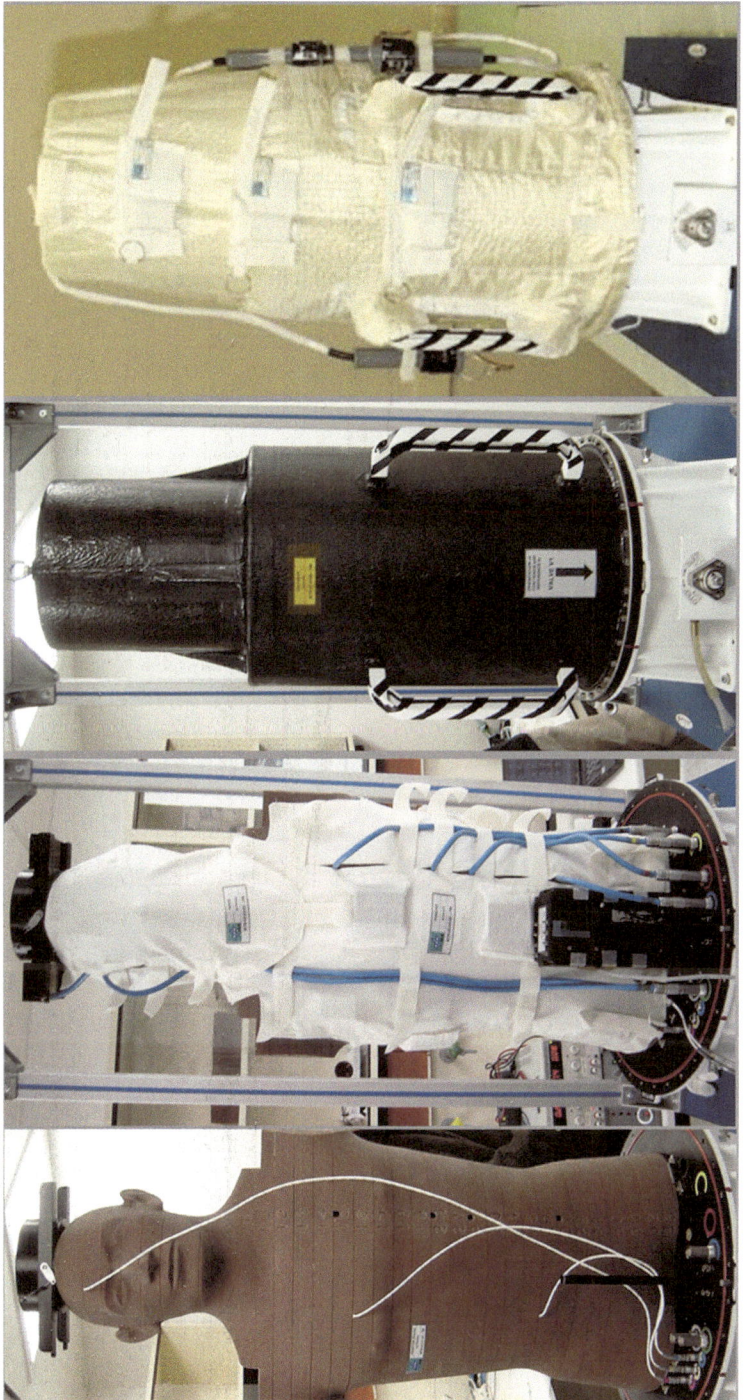

Figure 4.13: Human Phantom. This experiment has been around for years and is now in its umpteenth iteration. Radiation is a tough nut to crack when it comes to protecting astronauts. While astronauts on board ISS are not exposed to killer radiation doses (about 80 mSv during a 6-month increment, compared to 2 mSv during a year on Earth), when humans finally venture beyond Earth orbit (i.e: beyond the Moon – check your orbital mechanics for that statement) sometime in the 2040s or 2050s, it will be a different matter altogether. Osteoradionecrosis or cata-ractogenesis anyone? So, to better understand the radiation environment, the Human Phantom is used to collect data on radiation, thanks to hundreds of thermoluminescent detectors embedded at various depths in the torso. Credit: ESA

Table 4.6 Expedition 8

Mission	
Mission type:	ISS residency
Mission duration:	194 days, 18 hours, 33 minutes, 12 seconds
Orbits completed:	~3,170
Spacecraft	
Spacecraft:	Soyuz TMA-3 11F732
Crew	
Crew size:	3
Members:	C. Michael Foale
	Alexander Kaleri
	Pedro Duque (ESA, up only)
	André Kuipers (ESA, down only)
Callsign:	*Ingul*
EVAs	
Total EVAs:	1
Total duration:	3 hours, 55 minutes
Launch	
Launch date:	October 18, 2003, 05:38:03 UTC
Rocket:	Soyuz-FG
Launch site:	Baikonur 1/5
Landing	
Landing date:	April 30, 2004, 00:11:15 UTC
Landing site:	50.38° N, 67.20° E
Orbital parameters	
Perigee:	193 kilometers
Apogee:	227 kilometers
Inclination:	51.67 degrees

Expedition 8's sole spacewalk took place on February 26, 2004, less than two months before the end of the mission. In addition to installing Mr. Rando, Foale and Kaleri, wearing Russian Orlan suits, replaced a number of parts for existing experiments. Unfortunately, the duo's work was cut short due to a malfunction in Kaleri's suit, and their excursion outside the ISS lasted only three hours 55 minutes. Unlike previous ISS spacewalks, there were no crewmembers inside the station as the two-man 'caretaker' crew worked outside. The ingress of the Expedition 8 crew (Table 4.6) into ISS marked the end of Foale's spacewalking career. On April 21, the Expedition 9 crew was delivered by Soyuz TMA-4, and station command was handed over from Foale to Gennadi Padalka, who was accompanied by Mike Fincke and ESA astronaut Andre Kuipers. On April 29, Foale, Kaleri and Kuipers departed the ISS and headed home. While Kaleri would return to command the ISS in 2010 during Expedition 25/26 and Kuipers would fly to the outpost on Expedition 30/31, for Foale, the landing of Soyuz TMA-3 on the Kazakh steppe marked the end of a very illustrious six-flight astronaut career. At the end of the flight, Foale had logged over 374 days 11 hours in space on his six flights, making him the most experienced American (and, thanks to his dual nationality, British) astronaut. After receiving the Yuri Gagarin Gold Medal and a Commander of the Order of the British Empire from the Queen, Foale eventually retired from NASA in August 2013. He now works in the commercial sector, developing electric aircraft.

Figure 4.14: Mark Shuttleworth. Credit: Russian Space Agency

MARK SHUTTLEWORTH

A thoroughbred internet tycoon, Mark Shuttleworth made his fortune by establishing his company, Thawte, as a world leader in cyber security. He founded Thawte while still a student at the University of Cape Town in 1995. Four years later, he sold the company to VeriSign for $575 million – more than enough money to buy a ticket to the ISS. In 2001, Shuttleworth paid $20 million for his ticket to orbit and kick-started the 'First African in Space' venture. On April 25, 2002, after training for a year in Star City, Shuttleworth launched on Soyuz TM-34 (Figure 4.15), together with Commander Yuri Gidzenko and ESA flight engineer Roberto Vittori. He spent eight days on the ISS before returning on Soyuz TM-33 on May 5, 2002.

Shuttleworth holds dual South African and British nationality, a fact that was not lost on the UK media during his only spaceflight. The inspiration for Shuttleworth's mission was the controversial flight of the world's first 'space tourist', Dennis Tito[6], who had flown

---

[6] When Dennis Tito turned up at JSC for training, NASA Manager Bob Cabana told him to go home, saying "We will not be able to begin training, because we are not willing to train with Dennis Tito."

Figure 4.15: TM-34 mission patch. Credit: Russian Space Agency

to the ISS in 2001. Unlike Tito, Shuttleworth didn't simply want to be a passenger along for the ride and in the way, which is why he developed the First African in Space project, featuring a varied program of science projects that was shared with pupils across South Africa.

One of his investigations was the Embryo and Stem Cell Development Project, which sought to investigate the optimum growing environment for stem cells. By flying stem cells in space, scientists hoped to determine whether the zero-gravity environment was beneficial to cell growth. In the field of space-based developmental medicine, this question is significant because one day fetal growth will occur in zero gravity, and when that happens it will be important to know if the lack of gravity will have any detrimental effects.

Another experiment (*The effect of a microgravity environment on autonomic cardiovascular control, energy expenditure and muscle characteristics*) investigated the role of the autonomic nervous system on blood pressure and heart rate. This study sought to determine the relative contributions of the parasympathetic and sympathetic nervous system on Earth and in space during rest and exercise, to see if there was any change in heart rate variability between the two conditions.

A third investigation studied energy balance and expenditure by means of heart rate monitoring. This investigation was of particular relevance to South Africa, which at the time had one of the highest incidences of chronic lifestyle diseases in the world.

Shuttleworth's mission timeline

Day 1
Docking and boarding of ISS
Scheduled TV report

Day 2
Collect saliva swabs for CCE-2 experiment
Conduct cycling and rest tests for CCE-1 experiment
Ocean-monitoring phase of Plankton Lensa experiment using onboard camera

Stem-cell experiment
HAM radio session

Day 3
Perform expanders and cycle test for CCE-2 experiment
Ocean-monitoring phase of Plankton Lensa experiment using onboard camera

Day 4
Conduct cycling and rest test for CCE-1 experiment
Stem-cell experiment
Educational experiments
Ocean-monitoring phase of Plankton Lensa experiment
HAM radio session

Day 5
Collect second and final saliva swabs for CCE-2 experiment
Perform expanders and cycle test for CCE-2 experiment
HAM radio session
Ocean-monitoring phase of Plankton Lensa experiment

Day 6
Perform cycle and rest test for CCE-1 experiment
HAM radio session
Earth-surface monitoring phase of Plankton Lensa experiment
Stem-cell experiment

Day 7
Study landing timeline
Earth-surface monitoring phase of Plankton Lensa experiment
Ocean-monitoring phase of Plankton Lensa experiment
HAM radio session
Prepare return equipment

Day 8
Prepare return equipment and place isolated containers on standby
Earth-surface monitoring phase of Plankton Lensa experiment
HAM radio session
Deactivate crystals on SPC experiment and stow on return vehicle
Complete ESCD experiment and stow on return vehicle

Day 9
Finalize onboard experiments and load results onto return Soyuz

After retiring from his short-lived career as a cosmonaut, Shuttleworth became a philanthropist dedicated to improving access to education in South Africa, via his company HBD (Here Be Dragons) Venture Capital. He currently lives in a botanical garden on the Isle of Man, together with 18 ducks and his wife Claire.

Figure 4.16:   Piers Sellers. Credit: NASA

PIERS SELLERS

A meteorologist by training, Piers Sellers was born in Crowborough, Sussex, and went to school in Seaford, East Sussex, before earning his first degree at the University of Edinburgh and then a doctorate from the University of Leeds. He left the UK for the U.S. in 1982 and began his NASA career as a research meteorologist. In 1984, he applied for the astronaut corps, but was ineligible because he did not yet have U.S. citizenship. That problem was resolved in 1991 when he became a naturalized U.S. citizen, and Sellers applied again and was accepted as an astronaut candidate in April 1996. After completing astronaut training, he was assigned to the Computer Support Branch in the Astronaut Office before being called up for mission duty to STS-112, an ISS assembly flight (9A) that was to deliver the S1 Truss to the orbiting outpost.

First flight

Atlantis launched on October 7, 2002, a launch notable for the shedding of a chunk of foam from the External Tank's bipod ramp that caused a sizable dent near the base of the left Solid Rocket Booster (SRB)[7]. On the second flight day, Sellers and his crewmates

---

[7] Before STS-113, NASA managers had examined this incident and decided to proceed with launches as scheduled. It proved to be a fateful decision because the following mission – STS-107 – ended in disaster with the disintegration of *Columbia* on re-entry due to foam strike damage.

Figure 4.17: STS-112 mission patch. Credit: NASA

performed checkouts of the Canadarm, their spacesuits and the rendezvous tools in preparation for docking with the ISS. Two days of periodic engine firings brought *Atlantis* to a point just 15 kilometers behind the ISS. After firing its RCS, the Shuttle commenced its Terminal Intercept burn that marked the final rendezvous phase. As the Orbiter's rendezvous radar system began tracking the ISS, Commander Jeffrey Ashby and Pilot Pamela Melroy slowed the approach until *Atlantis* was moving at only 150 meters per hour just 100 meters from the station. After successfully docking, the crew began seven days of joint operations with the resident ISS crew – Expedition 5 Commander Valeri Korzun and Flight Engineers Peggy Whitson and Sergei Treschev.

The first spacewalk of STS-112 (Table 4.7) was performed by Sellers and David Wolf on October 10. During their seven-hour excursion outside the ISS, the duo attached power and data connections linking the S1 Truss and the first truss segment (S0) and uncoupled various power, video and data cables from positions on the S0 Truss. Two days later, Sellers and Wolf embarked on their second spacewalk. Tasked with bringing the S1 Truss to life, Sellers and Wolf spent their six hours outside connecting ammonia cooling system lines and various umbilicals to the Ammonia Tank Assembly (ATA) and installing cameras on the truss assembly. Then, three days later, the pair headed outside again to complete the checkout of the newly installed S1 Truss. Among their checkout tasks was the removal and replacement of the Interface Umbilical Assembly (IUA), a key element of the station's Mobile Transporter railcar that serves as a base for the Canadarm. Once that had been completed, Sellers and Wolf installed fluid jumpers to enable the flow of ammonia coolant between truss segments, before completing their final task which was a test of the Segment-to-Segment Attachment System (SSAS). The SSAS, which basically comprises claw and bolt assemblies, had to be tested to ensure it was ready to receive upcoming truss components on future flights. With their tasks complete, Sellers and Wolf headed back inside after six hours and 36 minutes and began preparing for the trip home.

Table 4.7 STS-112

Mission	
Mission type:	ISS assembly
Mission duration:	10 days, 19 hours, 58 minutes, 44 seconds
Distance Travelled:	7,200,000 kilometers
Orbits completed:	170
Spacecraft	
Spacecraft:	Shuttle *Atlantis*
Launch mass:	116,538 kilograms
Landing mass:	91,390 kilograms
Payload mass:	12,572 kilograms
Crew	
Crew size:	6
Members:	Jeffrey S. Ashby
	Pamela A. Melroy
	Piers J. Sellers
	Sandra H. Magnus
	David A. Wolf
	Fyodor N. Yurchikin (Russian)
EVAs	
Total EVAs:	3
Total duration:	19 hours, 41 minutes
Launch	
Launch date:	October 7, 2002, 19:45:51 UTC
Launch site:	Kennedy LC-39B
Landing	
Landing date:	October 18, 2002, 15:44:35 UTC
Landing site:	Kennedy SLF Runway 33
Docking with ISS	
Docking port:	PMA-2 (Destiny forward)
Docking date:	October 9, 2002, 15:16 UTC
Undocking date:	October 16, 2002, 15:13 UTC
Time docked:	6 days, 21 hours, 57 minutes
Orbital parameters	
Perigee:	273 kilometers
Apogee:	405 kilometers
Inclination:	51.6 degrees
Period:	91.2 min

Return-to-Flight

The next opportunity for Sellers to add to his time wearing a spacesuit came when he was assigned to the STS-121 mission on board *Discovery* (Figure 4.18). The main mission objective of STS-121 was to test the safety and repair techniques that had been implemented following the *Columbia* tragedy in February 2003, but the flight was also a cargo and assembly mission to the ISS (ULF 1.1). Launched on July 4, 2006, STS-121 was the only Shuttle flight that ever launched on Independence Day. Since STS-121 followed on

Figure 4.18:   STS-121 mission patch. Credit: NASA

from STS-114, the first Return-to-Flight mission, *Discovery's* flight was also categorized as a Return-to-Flight test, and it was only after *Discovery* returned that NASA resumed regular Shuttle flights. Some of the safety techniques involved inspection routines for the Shuttle's thermal protection system (TPS) and tests of methods to protect the TPS. In addition to its role in testing safety techniques, STS-121 (Table 4.8) was notable for returning the ISS crew strength to three, as it transported ESA astronaut Thomas Reiter to the ISS to join Expedition 13. Given the 'safety' emphasis of the mission, it is not surprising that this Shuttle mission became the most photographed of the entire program.

As with all assembly missions, spacewalks featured prominently on STS-121, which had three of them scheduled during the 13-day stay on orbit. They were all conducted by the pairing of Sellers and Michael Fossum. The first of these, which was the fourth spacewalk for Sellers and the first for Fossum, saw the duo reroute and replace cable around the S0 Truss, in addition to testing the Canadarm as a platform for astronauts to repair a damaged Shuttle if such an event occurred. Two days later, on Flight Day 7, the pair were in action again, this time restoring the station's Mobile Transporter railcar to full operations and replacing umbilical cables. At six hours 47 minutes, this second spacewalk was a shorter affair than the seven hours and 31 minutes of the first and the seven hours and 11 minutes of the third, which took place on Flight Day 9. Their third spacewalk (Figure 4.19) should have been a very routine excursion (or as routine as EVAs get), because it was basically a nuts-and-bolts repair demonstration of a pre-ceramic polymer sealant that had been developed to be used on damaged TPS (carbon-carbon) panels. The spacewalk wouldn't have attracted much attention had it not been for Sellers losing a spatula that was used to put the sealant onto the panels. Sellers apologized to the ground, who told the embarrassed astronaut not to worry. While it was the first spatula lost in space, two years later it had company when Heidemarie Stefanyshyn-Piper lost an entire tool bag (valued at more than $100,000) that contained two such spatulas. Happily, items lost from the altitudes that Shuttles flew generally don't pose a debris hazard, as evidenced by a note that NORAD sent to Sellers four months after the incident, informing him of where the spatula had fallen back to Earth.

Table 4.8 STS-121

Mission	
Mission type:	ISS logistics and Shuttle Return-to-Flight
Mission duration:	12 days, 18 hours, 37 minutes, 54 seconds
Distance Travelled:	8,500,000 kilometers
Orbits completed:	202
Spacecraft	
Spacecraft:	Shuttle *Discovery*
Launch mass:	121,092 kilograms
Payload mass:	14,594 kilograms
Crew	
Crew size:	7 up, 6 down
Members:	Steven W. Lindsey
	Mark E. Kelly
	Michael E. Fossum
	Lisa M. Nowak
	Stefanie D. Wilson
	Piers J. Sellers
	Thomas A. Reiter (ESA, up only)
EVAs	
Total EVAs:	3
Total duration:	21 hours, 29 minutes
Launch	
Launch date:	July 4, 2006, 18:37:55 UTC
Launch site:	Kennedy LC-39B
Landing	
Landing date:	July 14, 2006, 13:14:43 UTC
Landing site:	Kennedy SLF Runway 15
Docking with ISS	
Docking port:	PMA-2 (Destiny forward)
Docking date:	July 6, 2006, 14:52 UTC
Undocking date:	July 15, 2006, 10:08 UTC
Time docked:	8 days, 19 hours, 16 minutes
Orbital parameters	
Perigee:	352.8 kilometers
Apogee:	354.2 kilometers
Inclination:	51.6 degrees
Period:	91.6 min

Final flight

Four years later, Sellers was back in action for STS-132 (Figure 4.20), another ISS assembly mission (ULF4) that launched on May 14, 2010. On board *Atlantis* was the Russian Rassvet Mini Research Module (MRM) and the Integrated Cargo Carrier-Vertical Light Deployable (ICC-VLD), a modular pallet that had previously flown on STS-127. STS-132

Figure 4.19: Piers Sellers translates along a truss during the third EVA of STS-121. Credit: NASA

Figure 4.20: STS-132 mission patch. Credit: NASA

was notable because it was slated to be the final flight of *Atlantis*, assuming the STS-135 Launch-on-Need (rescue) mission would not be required. But in February 2011, NASA decided to add one final mission, and *Atlantis* flew once again for the STS-135 flight in July 2011. At the time, however, people assumed they were witnessing the final flight of *Atlantis* with STS-132[8], and unsurprisingly the media coverage and celebrity attendance (including Buzz Aldrin and David Letterman) was higher than it had been for previous flights.

On their first full day on orbit, the crew inspected the Orbiter's TPS and prepared their spacesuits in readiness for the planned EVAs. Meanwhile on the ground, Mission Control studied the path of a piece of orbital debris that appeared as if it might require the ISS to perform an avoidance maneuver. Fortunately, updated tracking data revealed that the debris would not present a hazard. On Flight Day 3, *Atlantis* continued its rendezvous burns to boost its orbit to match that of the ISS. After executing its 360° back-flip Rendezvous Pitch Maneuver to allow ISS crewmembers to photograph the Orbiter's underside, *Atlantis* docked with Pressurized Mating Adapter-2. After a brief welcome ceremony with the incumbent ISS crew of Oleg Kotov, Timothy Creamer and Soichi Noguchi, the crew transferred spacesuits into the ISS, and Tracy Caldwell-Dyson and Sellers got to work relocating the ICC-VLD, using the Canadarm to move the pallet from the cargo bay to the ISS Mobile Base System in preparation for the mission's first excursion outside.

As with all spacewalks, preparations began with resizing the suits, checking translation paths and the pre-breathe procedure. There may be some reading this who think that the astronauts just slip into the bulky suits, pop the airlock and head outside, but entering the vacuum of space, even while wearing a $12 million spacesuit, is an almost unfathomably complex exercise, as evidenced by the Shuttle era's 233-page EVA checklist (the ISS version is a whopping 644 pages! – see sidebar for a snapshot of just *one* procedure). To even touch on the complexities involved in sending humans into a vacuum is way beyond the scope of this book, so let's just take a look at the pre-breathe. Why do astronauts have to do this? Well, spacesuits work at pressures that are lower than the cabin atmosphere, and working at lower atmospheric pressures increases the risk of decompression sickness (DCS). The workaround for this is the procedure known as *pre-breathing*. DCS is caused by nitrogen bubbles forming in the blood vessels and moving to other areas of the body where they can cause damage, so the astronauts engage in the pre-breathe protocol designed to wash out nitrogen from the body. Since this procedure takes many hours, astronauts perform exercises to accelerate the rate at which nitrogen is flushed from the body. Even with this exercise, the procedure still takes over two hours. Essentially, the ISS protocol (dubbed ISLE, for In-Suit Light Exercise) is as follows:

- Astronauts exercise at high intensity on the ISS ergometer for 10 minutes, breathing pure oxygen from an oxygen mask.

---

[8] Because the flight was planned as the last, the STS-132 mission patch featured *Atlantis* flying into the sunset and into retirement. On her return from STS-132, *Atlantis* was readied as a rescue Shuttle for the last missions of *Discovery* (STS-133) and *Endeavour* (STS-134).

- They spend another 40 minutes breathing oxygen.
- The pressure in the airlock is reduced to 10.2 psi. During airlock depressurization, the astronauts continue to breathe oxygen for 30 minutes.
- With the airlock depressurized, the astronauts don their suits.
- With their suits on, the astronauts breathe oxygen for another 60 minutes.

Procedure for Failed Leak Check
FAILED LEAK CHECK (14.7/10.2 PSI)
NOTE
The following steps are performed on the leaking EMU only
1. Leaking EMU: O2 ACT – PRESS IV
2. Rotate lower arm assemblies 180 degrees cw and 360 degrees ccw
3. Align suit arms
4. √Sizing rings locked
5. Swivel hips from side to side
6. Repeat leak check as follows:
√Helmet purge vlv – cl, locked
DCM √PURGE vlv – cl (dn)
√O2 ACT – PRESS until SUIT P = 4.2–4.4 and
stable (compare w/gauge)
O2 ACT – IV, start timing, 1 min
(during EMU CHECKOUT, 2 min)
(Max $\Delta P = 0.3$ psi)
7. If leak check passed, go to step 11
If leak check failed:
DCM PURGE vlv – op (up), O2 ACT – OFF
FAN – OFF (if EVA PREP)
Cycle/inspect suit disconnects as follows:
Gloves, helmet (leave off), LTA, boots (if removed in EVA PREP) FAN – ON
(if EVA PREP)
8. Install helmet, repeat leak check step 6, then:
9. If leak check passed, go to step 11
   If leak check failed:
DCM PURGE vlv – op (up), O2 ACT – OFF
FAN – OFF (if EVA PREP)
Cycle/inspect suit disconnects as follows:
Gloves, helmet (leave off)
Helmet purge vlv, Sizing rings
LiOH cartridge (O-rings)

FAN – ON (if EVA PREP)
10. Install helmet, repeat leak check step 6, then:
11. If leak check passed:
√Waist ring, wrist rings covered
Continue EMU CHECKOUT or
EVA PREP >>
12. If leak check failed (EMU lost): Contact MCC

For Sellers, who had performed three spacewalks on both of his previous missions, watching Garrett Reisman and Stephen Bowen suit up for spacewalk operations on Flight Day 4 must have felt a little strange. However, in his role as Canadarm operator (together with Caldwell-Dyson), Sellers still played an integral role in the EVA operations. One of Reisman and Bowen's first tasks was to install a new tool platform for Dextre, aka the Canadarm. Once that had been completed, they moved on to the installation of a new Space to Ground ANTenna (SGANT), a job that presented more than its fair share of problems. The first challenge was to close a gap between the antenna dish and its mounting pole, a snag that was resolved by cranking up the torque setting on the bolts. Then the command and control computer detected an error and shut down, forcing more trouble-shooting. Their jobs done, the duo headed back to the Quest airlock, closing out the 144th spacewalk in support of ISS assembly operations.

The following day, both crews gathered in the Harmony module for a media get together with MSNBC, Fox News and CNN, to answer questions about life on orbit, their spacewalking experiences and the myriad science experiments being conducted on ISS. Bowen and Michael Good then headed off to the airlock to begin their camp-out in preparation for the mission's second spacewalk the following day. The main objective of EVA 2 was to remove and replace batteries on the P6 Truss, a task choreographed by pilot Tony Antonelli and assisted by Commander Ken Ham, who provided video support. The following day – Flight Day 7 – kept the crew busy with more spacewalk prepara-tions, while ISS crewmembers Kotov and Aleksandr Skvortsov checked out the MRM module. EVA 3 was performed by Reisman and Good, who connected ammonia jumpers on the P4 and P5 Truss segment and swapped out batteries on the P6 Truss during their time outside.

With spacewalking duties completed, the *Atlantis* crew kicked back and enjoyed some off-duty time on Flight Day 9 before turning their attention to the ICC-VLD, which Sellers and Reisman returned to the bay using the Canadarm. The next day, *Atlantis* undocked from ISS, performed two separation burns, and edged away from the station. Flight Day 11 was devoted to inspecting the Shuttle's leading edges and nose using the Canadarm, and stowing the spacesuits in preparation for pre-landing activities the following day. On Flight Day 12, the *Atlantis* crew conducted hot-fire checkouts, stowed cargo in preparation for landing and finished the day chatting with Stephen Colbert. STS-132 (Table 4.9) came to end on Flight Day 13 when *Atlantis* landed on Runway 33 at the Kennedy Space Center.

Table 4.9 STS-132

Mission	
Mission type:	ISS logistics and assembly
Mission duration:	11 days, 18 hours, 29 minutes, 9 seconds
Distance Travelled:	7,853,563 kilometers
Orbits completed:	186
Spacecraft	
Spacecraft:	Shuttle *Atlantis*
Launch mass:	2,050,133 kilograms (total), 119,300 kilograms (orbiter)
Landing mass:	95,024 kilograms
Payload mass:	12,072 kilograms
Crew	
Crew size:	6
Members:	Kenneth T. Ham
	Dominic A "Tony" Antonelli
	Garrett E. Reisman
	Michael T. Good
	Stephen G. Bowen
	Piers J. Sellers
EVAs	
Total EVAs:	3
Total duration:	21 hours, 20 minutes
Launch	
Launch date:	May 14, 2010, 18:20 UTC
Launch site:	Kennedy LC-39A
Landing	
Landing date:	May 26, 2010, 12:49:18 UTC
Landing site:	Kennedy SLF Runway 33
Docking with ISS	
Docking port:	PMA-2 (Harmony forward)
Docking date:	May 16, 2010, 14:28 UTC
Undocking date:	May 23, 2010, 15:22 UTC
Time docked:	7 days, 1 hour, 1 minute
Orbital parameters	
Perigee:	335 kilometers
Apogee:	359 kilometers
Inclination:	51.6 degrees
Period:	91 min

For Sellers, STS-132 marked the end of his astronaut career. The following year, he took up the position of Deputy Director of Sciences and Exploration at NASA's Goddard Space Center. Four years later, in October 2015, he was diagnosed with Stage 4 pancreatic cancer. He died on December 23, 2016.

"When I was a kid, I watched the Apollo launches from across the ocean, and I thought NASA was the holy mountain. As soon as I could, I came over here to see if I could climb that mountain. I used to cut out all the magazines and collect all the pictures; I thought the whole thing was wonderful. So I thought, 'Wow, I'd love to be an astronaut. I talked to my teachers about my desire to be an astronaut, and they said, 'Well, you should do science.' So off I went into science, and I found out I thoroughly enjoyed it."

Piers Sellers, on being awarded NASA's Distinguished Service Medal, the agency's highest honor, in June 2016. National Geographic.

"I've no complaints. As an astronaut, I spacewalked 220 miles above the Earth. Floating alongside the International Space Station, I watched hurricanes cartwheel across oceans, the Amazon snake its way to the sea through a brilliant green carpet of forest, and gigantic nighttime thunderstorms flash and flare for hundreds of miles along the Equator. From this God's-eye-view, I saw how fragile and infinitely precious the Earth is. I'm hopeful for its future."

Piers Sellers, writing in The Times following his cancer diagnosis.

NICHOLAS PATRICK

The fourth person born in the UK to visit space is Nicholas Patrick. He completed the first part of his university education in the UK, gaining an undergraduate degree in engineering from Trinity College, Cambridge, in 1986, before heading off to MIT, where he received a PhD in 1996. His opportunity to become an astronaut became a reality in 1994 when he became a U.S. citizen, a status he took advantage of two years later when he was selected to NASA's Group 17. After completing basic astronaut training, Patrick was selected to be an aquanaut on board the Aquarius underwater laboratory (see sidebar) for the NEEMO 6 mission in July 2004.

Aquarius
Aquarius is a subsea lab that sits in 20 meters of water a few kilometers offshore from Key Largo, Florida. Operated by Florida International University, Aquarius has hosted several NEEMO (NASA Extreme Environment Mission Operations) missions over the years because the environment provides a setting similar to that experienced by astronauts on orbit. Aquanauts living in the lab do so by employing saturation diving techniques, which means that at the end of a typical 10-day mission, they must spend more than 15 hours decompressing. The system itself comprises a habitat module that measures more than 11 meters in length and three meters in diameter, a Main Lock that features berths for the crew, and an Entry Lock that contains computers and life support controls. On the surface is a support buoy tethered to a five-point mooring. The 10-meter diameter buoy contains generators, compressors and telemetry and control systems that transmit data to mission control situated in Key Largo.

Figure 4.21: Nicholas Patrick. Credit: NASA

The sixth NEEMO mission, which took place between July 12–21, 2004, was dedicated to spaceflight biomedical research. The crew comprised Patrick, Doug Wheelock, John Herrington, and Tara Ruttley. During their time in the underwater habitat, the crew evaluated medical system concepts, tested physiological monitoring systems, and conducted medical experiments such as the Portable Bone Quality Assessment Device.

Two years later, Patrick found himself ensconced inside *Discovery*, waiting to be launched to the ISS. In addition to delivering the station's $11 million girder-like P5 Truss segment[9], the STS-116 mission (Figure 4.22) was tasked with carrying a Spacehab Logistics Module and exchanging ISS Expedition 14 crewmembers: STS-116 crewmember Sunita Williams would be replacing ISS Flight Engineer Thomas Reiter, who would join the STS-116 crew for the return to Earth.

---

[9] The ISS Trusses form the backbone of the station. Some, like the P5 Truss (which measures 3.3 by 4.5 by 3.2 meters and weighs 1,800 kilograms), serve as a mounting location for logistics carriers and hardware, in addition to connecting power and cooling lines between other trusses (there are six port-side trusses and six starboard trusses). The predominantly aluminum structure also provides robotic interfaces and extravehicular translation aids.

Figure 4.22: STS-116 mission patch. Credit: NASA

A busy schedule

Discovery launched on December 9, 2006, marking the third Shuttle mission in five months (following STS-121 in July and STS-115 in September). As was becoming routine, the first task on orbit was a thorough inspection of the Shuttle using cameras mounted on the Canadarm. Patrick's job during the five-plus hour evaluation was to inspect the leading edge of the wings and the nose cap. With inspection tasks complete, *Discovery* headed towards the ISS and completed the now-familiar nine-minute back-flip Rendezvous Pitch Maneuver to allow station crewmembers to snap images of the Orbiter's heat shield from the windows of the Zvezda Service Module[10]. Docking took place on Flight Day 3, with Patrick playing a role by processing navigation data and operating the laser range systems, together with Mission Specialists William Oefelein and Joan Higginbotham. Later, he helped Higginbotham and Williams operate the Canadarm to prepare for the first of the mission's three spacewalks, by Robert Curbeam and Christer Fuglesang.

The two EVA astronauts headed out of the airlock on December 12, with a task list that included aligning and connecting the P5 Truss to the P4 Truss. The duo also replaced a video camera attached to the S1 Truss before heading back inside the station. Two days later, the pair headed out once more, this time to reconfigure power for the station's electrical system. This required a lengthy power down procedure that was executed from the ground while Curbeam and Fuglesang prepared for the excursion outside. Once outside, the pair of spacewalkers got to work on the station's main bus (Unit 2) switching unit (these route power from the solar arrays to the DC-to-DC converter units, which then convert 160-volt DC electricity to 125 volts). While Curbeam wrestled with Unit 2, Fuglesang worked on the Circuit Interrupt Devices (CIDS), the station's version of a circuit breaker. With that job checked off, the pair went to work on other maintenance tasks, including

---

[10] The ISS crews used 400-mm and 800-mm digital camera lenses during these inspections. The 400-mm lens provided 7.6 cm resolution, while the 800-mm lens provided 2.5 cm resolution. The images taken included those of the elevon cover areas, landing gear door seals and the Orbiter's underside.

installing thermal covers on the force moment sensors on the Canadarm's latching end effectors, and reconfiguring power to the station's Z1 Truss electrical panel. As the spacewalk edged past the five-hour mark, Curbeam and Fuglesang headed back inside the station to rest up before their third and final work increment outside.

That operation began two days later, when Curbeam and Williams conducted a spacewalk that was very similar to the electrical reconfiguration tasking from a few days previously. On paper, the EVA looked fairly straightforward, but when working in a vacuum, there's no such thing as straightforward, just variations of 'complicated', and so it proved during the third EVA. After completing their wiring tasks, Curbeam and Williams installed a Canadarm grapple fixture and placed a package of debris shields outside the service module, before attempting to retract the balky P6 solar array. They were able to achieve 65 percent retraction before being called back in after spending seven hours and 31 minutes outside. With the P6 solar array only partially retracted, a fourth spacewalk was hastily planned, and Curbeam and Fuglesang headed out once more to fully retract the balky array. For Curbeam, his fourth EVA on STS-116 (Table 4.10) set a record for the most spacewalks during a Shuttle mission, his total topping out at 25 hours and 45 minutes.

Table 4.10 STS-116

Mission	
Mission type:	ISS assembly and resident crew exchange
Mission duration:	12 days, 20 hours, 44 minutes, 16 seconds
Distance Travelled:	8,500,000 kilometers
Orbits completed:	203
Spacecraft	
Spacecraft:	Shuttle *Discovery*
Launch mass:	120,413 kilograms
Landing mass:	102,220 kilograms
Payload mass:	12,523 kilograms
Crew	
Crew size:	7 (including exchange)
Members:	Mark L. Polansky
	William A. Oefelein
	Nicholas J.M. Patrick
	Robert L. Curbeam Jr.
	Christer Fugelsang (ESA)
	Joan E. Higginbotham
	Sunita L. Williams (up only)
	Thomas A. Reiter (down only)
EVAs	
Total EVAs:	4
Total duration:	25 hours, 45 minutes
Launch	
Launch date:	December 9, 2006, 01:47:35 UTC
Launch site:	Kennedy LC-39B

(continued)

Table 4.10 (continued)

Landing	
Landing date:	December 22, 2006, 22:32:00 UTC
Landing site:	Kennedy SLF Runway 15
Docking with ISS	
Docking port:	PMA-2 (Destiny forward)
Docking date:	December 11, 2006, 22:12 UTC
Undocking date:	December 19, 2006, 22:10 UTC
Time docked:	7 days, 23 hours, 58 minutes
Orbital parameters	
Perigee:	326 kilometers
Apogee:	358 kilometers
Inclination:	51.6 degrees
Period:	91.37 min

With assembly tasks complete, the STS-116 crew prepared to go home. *Discovery's* heat shield was inspected and the Orbiter separated from the ISS. Oefelein performed a fly-around of the station before firing the Shuttle's jets and carefully guiding *Discovery* away. Originally, the landing had been planned for December 21, but due to the fourth spacewalk the landing was pushed back to December 22, a day that turned out to be less than optimal for landing a Shuttle. Returning to Edwards Air Force Base was a no-go because of high cross-winds, while clouds and showers initially precluded a landing at KSC. The remaining option was White Sands Space Harbor, which had last seen a landing in 1982 when the STS-3 mission returned there. Fortunately, the weather at KSC abated, and *Discovery* was able to touch down shortly before dusk. It was the 64th landing at KSC.

Rearranging the furniture

Patrick's next mission was STS-130 (Figure 4.23) on board *Endeavour*. Tasked with the usual shopping list of jobs, the STS-130 crew was assigned to activate and checkout the Tranquility module, connect myriad avionics cables and ammonia cooling jumpers, and relocate the Cupola from the end port of the module to the Earth-facing port (see sidebar) and reposition the Pressurized Mating Adapter 3 (PMA-3) from its location on Harmony to the port berth on Tranquility

The Cupola

The Cupola (also referred to by the crew as the Millennium Falcon for obvious reasons - see Figure 4.24) is a dome-shaped single-forged aluminum structure that features six trapezoidal side windows and one circular Earth-facing window 80 centimeters in diameter. The windows are made of bullet-proof, super-tough fused silica glass panes – protected by external shutters. Just in case. Functioning as the outpost's panoramic control tower, the Cupola doubles as a place of recuperation for the astronauts and as a workplace from which crewmembers can manipulate the Canadarm using robotics workstations.

Figure 4.23: STS-130 mission patch. Credit: NASA

Figure 4.24: Nicholas Patrick looks out through the Cupola windows. Credit: NASA

Endeavour's STS-130 mission (Table 4.11) launched on February 8, 2010 and followed a familiar trajectory to rendezvous with the ISS. During Flight Day 2, much of which was spent conducting the inspection of the TPS, Patrick and Bob Behnken got to work preparing the spacesuits they would be using during the mission's three EVAs. Docking with the ISS occurred on Flight Day 3. After the familiar welcome and safety briefing, George Zamka, Behnken and Steve Robinson transferred the spacesuits from the Shuttle to the ISS in preparation for the following day's pre-EVA camp-out in the airlock.

For Patrick and Behnken, much of Flight Day 4 was spent preparing for their spacewalk the following day. Tools were sorted (see Appendix II for a description of some of the

Table 4.11 STS-130

Mission	
Mission type:	ISS assembly
Mission duration:	13 days, 18 hours, 6 minutes, 22 seconds
Distance Travelled:	9,250,000 kilometers
Orbits completed:	217
Spacecraft	
Spacecraft:	Shuttle *Endeavour*
Launch mass:	121,320 kilograms
Landing mass:	91,033 kilograms
Payload mass:	15,844 kilograms
Crew	
Crew size:	6
Members:	George D. Zamka
	Terry W. Virts Jr.
	Kathryn P. Hire
	Stephen K. Robinson
	Nicholas J.M. Patrick
	Robert L. Behnken
EVAs	
Total EVAs:	3
Total duration:	18 hours, 14 minutes
Launch	
Launch date:	February 8, 2010, 09:14:07 UTC
Launch site:	Kennedy LC-39A
Landing	
Landing date:	February 22, 2010, 03:20:29 UTC
Landing site:	Kennedy SLF Runway 15
Docking with ISS	
Docking port:	PMA-2 (Destiny forward)
Docking date:	February 10, 2010, 05:06 UTC
Undocking date:	February 20, 2010, 00:54 UTC
Time docked:	9 days, 19 hours, 47 minutes
Orbital parameters	
Perigee:	334 kilometers
Apogee:	348 kilometers
Inclination:	51.6 degrees
Period:	91.4 min

tools used by astronauts during EVAs), Hard Upper Torsos (HUT) checked, video systems tested, heaters in the gloves and boots examined, and the myriad list of pre-EVA activities ticked off one by one. Then, after dinner, the crew conducted a pre-EVA review and it was time for Patrick and Behnken to retire to the Quest Airlock to begin the camp-out phase of their pre-EVA pre-breathe.

The next day was the first of three scheduled EVA days, with Patrick (Figure 4.25) and Behnken starting procedures by preparing the release launch locks on the Tranquility Module

Figure 4.25:   Nicholas Patrick during EVA on STS-130. Credit: NASA

and the Cupola. Tranquility was then moved to the port side of Unity using the Canadarm. With the new addition to the ISS in place, Patrick and Behnken connected temporary heater and data cables between Unity and Tranquility before returning to the airlock having spent six-and-a-half hours outside. With Tranquility in place, it was time to perform initial outfitting. This task began on Flight Day 6. While Terry Virts, Stephen Robinson and Kay Hire were busy with the outfitting tasks, Patrick and Behnken prepared for their second spacewalk, scheduled for the next day. EVA 2 lasted 5 hours 54 minutes on Flight Day 7, much of which was spent installing ammonia cooling loops, thermal blankets (to protect ammonia hoses) and handrails, and outfitting Tranquility's Earth-facing port in preparation for the Cupola, which was to be moved to its permanent location the following day.

The highlight of Flight Day 8 was the relocation of the Cupola from its launch location to its permanent position on the nadir (Earth) facing side of Tranquility, a task that was executed by the Canadarm operated by Hire and Virts. Once the Cupola had been positioned, Jeff Williams secured the viewing port by tightening bolts around the structure in preparation for crew ingress the following day. While the Cupola activities were going on, Patrick and Behnken prepared equipment for their final walk in space, scheduled for Flight Day 10. Before the mission's third EVA, the crew had more repositioning activities to complete, including relocating the Pressurized Mating Adapter 3 (PMA-3) from Harmony to the end of Tranquility. While Behnken, Patrick, Williams and ISS crewmember Soichi Noguchi performed this task, Hire and Virts continued outfitting the Cupola.

The third and final spacewalk of STS-130 lasted 5 hours 48 minutes and featured the connection of data and heater cables to PMA-3, removal of the thermal covers on the Cupola and installation of handrails on Tranquility. The following couple of days were devoted mainly to housekeeping: Patrick and Behnken tidied up the airlock for use by ISS crews, rack transfers were completed, and the station was reboosted using its vernier thrusters. Then it was time to head home. *Endeavour* undocked on Flight Day 13 and spent its last day on orbit (Flight Day 14) performing pre-landing checks before executing a landing at KSC.

Patrick retired from NASA in 2012 and has since worked for Blue Origin – he was the Launch Safety Officer for New Shepard missions 1 through 6, which included the first powered landing of a booster returning from space (this was not a SpaceX first, as many seem to think) and the capsule escape test conducted in October 2016.

GREGORY JOHNSON

Born in South Ruislip in London, Gregory H. Johnson (there have been two NASA astronauts called Greg Johnson; the other, Gregory C. Johnson, was born in the U.S.) followed a well-worn route to becoming a NASA astronaut by first serving in the military. After receiving his commission from the United States Air Force Academy in 1984, he qualified as a pilot, going on to serve as an instructor on the T-38A before being selected as a F-15E pilot. He flew 34 combat missions in Operation *Desert Storm*, and another 27 combat missions in support of Operation *Southern Watch*, before being selected for Test Pilot School at Edwards Air Force Base (EAFB) in 1993. After spending a few years at EAFB testing aircraft such as the F-15C/E and T-38A/B, Johnson was selected by NASA as an astronaut in 1998. After completing astronaut training, he was assigned to various positions in the Flight Crew Operations Directorate, and in support of STS-100 and STS-108 as chief of Shuttle abort planning. His first opportunity to fly to space was as pilot for STS-123 (Figure 4.27).

A bit of everything

STS-123 (ISS assembly flight 1J/A), was flown by *Endeavour* and launched on March 11, 2008. The 25th Shuttle mission to the ISS was tasked with delivering the Japanese Experiment Module (JEM) and the Canadian Special Purpose Dexterous Manipulator (SPDM, aka 'Dextre') to the station, in addition to conducting a busy schedule of five spacewalks. Following the standard inspection routine, docking occurred on Flight Day 3, and spacewalkers Richard Linnehan and Garrett Reisman began preparing their gear for the first of the EVAs which was scheduled for the following day. Flight Day 4 featured Linnehan and Reisman installing Dextre, as well as crewmembers on board *Endeavour* using the Canadarm to remove the Japanese Logistics Module (JLM) from the Shuttle's payload bay and attach it to the ISS. The following day was spent outfitting the JLM and transferring supplies from *Endeavour* to the module, while Linnehan and Mike Foreman spent their time preparing for the second EVA. During their 7-hour 8-minute spacewalk on Flight Day 6, Linnehan and Foreman completed the installation of Dextre while their

Figure 4.26: Gregory H. Johnson, who served as pilot on *Endeavour* missions STS-123 and STS-134, was born in the UK but never held British citizenship. In addition to Johnson and the other astronauts with links to Britain described in this chapter, it is worth mentioning those who came close to making the journey to space. Dr. Anthony Llewellyn was selected as a scientist-astronaut by NASA during August 1967 but resigned the following year without ever having flown in space. Army Lieutenants-Colonel Anthony Boyle and Richard Farrimond, MoD employee Christopher Holmes, Royal Navy Commander Peter Longhurst and RAF Squadron Leader Nigel Wood (all born in the UK) were selected in February 1984 as Payload Specialists for the military Skynet 4 program. Boyle resigned from the program in July 1984, while Wood was due to fly aboard STS-61-H in 1986 (with Farrimond as back-up), but the *Challenger* accident meant that Wood never flew. Longhurst was due to fly aboard STS-71C in 1987 (with Holmes as back-up), but both resigned in 1986 without flying. Credit: NASA

Figure 4.27: STS-123 mission patch. Credit: NASA

fellow crewmembers continued the job of outfitting the JLM, a task that continued the following day. Flight Day 8 featured EVA 3, conducted by Linnehan and Robert Behnken, who performed a 6-hour 53-minute excursion to install a space parts platform and calibrate Dextre's end effector. The following day was heavy on robotics activity, as Dextre was attached to a power and data grapple fixture and the Canadarm was used to return the pallet used to secure Dextre during launch. The preparations for the fourth EVA began late on Flight Day 10, as Behnken and Foreman retired to the airlock to begin the standard pre-EVA camp-out. The next day, the duo replaced a circuit breaker on the station's truss and tested a repair procedure for damaged Shuttle tiles (see sidebar).

NASA's 'Goo Gun'
The Tile Repair Ablator Dispenser (T-RAD) comprises a pressurized canister and hand-held caulk 'gun' that can be attached to an astronaut's suit. The device dispenses an ablator material known as Shuttle Tile Ablator-54 (STA-54) that acts as a sealant for door seals and penetrations.

The main events of Flight Day 12 were the final inspection of Endeavour's TPS using the Canadarm and the preparations for the mission's fifth spacewalk. During their six-hour EVA, Foreman and Behnken completed various maintenance and housekeeping chores, including inspecting a jammed rotary joint that had been restricting the use of one of the sets of solar arrays. The following day was spent preparing for departure and kicking back for some well-deserved off-duty time. STS-123 (Table 4.12) departed from the ISS on Flight Day 15, and the mission landed at KSC during Flight Day 17.

"*Endeavour's* always been my favorite vehicle, I guess, because it's the newest of the vehicles. When I first became an astronaut, one of my early jobs was to help prepare the vehicle and the crews for launch at the Cape; flipping switches, setting up procedures, taping things down. And so I became familiar with all of the different shuttles because I spent a lot of time in all of them. *Endeavour* always looked the cleanest, it was the most pristine, and it was my favorite vehicle from the very start."

Gregory Johnson, interviewed by NASA on his retirement

Endeavour's last flight

STS-134 (Figure 4.28), aka ULF6, was the final flight of *Endeavour* and the penultimate flight of the Shuttle program. Launched on May 16, 2011, *Endeavour*, piloted by Gregory Johnson and commanded by Mark Kelly (Figure 4.29), docked with the ISS on Flight Day 3. The first task item on the agenda was to remove the Express Logistics Carrier from *Endeavour's* payload bay and reposition it on the P3 Truss segment. With that job checked off, Mike Fincke and Drew Feustel got to work transferring the two EMUs to the station's airlock in preparation for the four planned EVAs. Flight Day 4 was highlighted by the installation of the AMS-2, which was removed from the payload bay by Johnson and

Table 4.12 STS-123

Mission	
Mission type:	ISS assembly and resident crew exchange
Mission duration:	15 days, 18 hours, 10 minutes, 54 seconds
Distance Travelled:	10,585,900 kilometers
Orbits completed:	250
Spacecraft	
Spacecraft:	Shuttle *Endeavour*
Launch mass:	122,364 kilograms
Landing mass:	94,158 kilograms
Payload mass:	16,916 kilograms
Crew	
Crew size:	7 (including crew exchange)
Members:	Dominic L. Gorie
	Gregory H. Johnson
	Robert L. Behnken
	Michael J. Foreman
	Richard M. Linnehan
	Takao Doi (JAXA)
	Garrett E. Reisman (up only)
	Leopold Eyharts (ESA, down only)
EVAs	
Total EVAs:	5
Total duration:	33 hours, 28 minutes
Launch	
Launch date:	March 11, 2008, 06:28:14 UTC
Launch site:	Kennedy LC-39A
Landing	
Landing date:	March 27, 2008, 00:40:41 UTC
Landing site:	Kennedy SLF Runway 15
Docking with ISS	
Docking port:	PMA-2 (Harmony forward)
Docking date:	March 13, 2008, 03:49 UTC
Undocking date:	March 26, 2008, 00:25 UTC
Time docked:	11 days, 20 hours, 36 minutes
Orbital parameters	
Perigee:	336 kilometers
Apogee:	346 kilometers
Inclination:	51.6 degrees
Period:	91.6 min

Greg Chamitoff using the Canadarm. Once the AMS-2 had been installed on the S3 Truss, Chamitoff and Feustel prepared their EMUs for the spacewalk scheduled for the following day.

Flight Day 5 featured a 6-hour 19-minute excursion, during which Feustel and Chamitoff installed ammonia jumpers between the P3 and P6 Trusses and installed a new light on the Crew Equipment Translation Aid (CETA). The following day once again featured operation

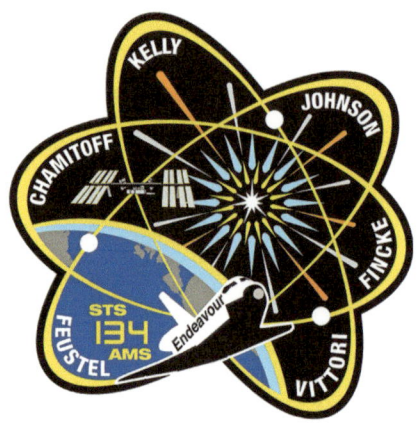

Figure 4.28:   STS-134 mission patch. Credit: NASA

Figure 4.29:   Mark Kelly and Greg Johnson, STS-134. Credit: NASA

of the Canadarm, controlled by Johnson, who used the robotic arm to inspect the Orbiter's TPS. With the inspection over, the crew took a time-out for an interview with Pope Benedict, who bestowed a blessing on the crew during the first-ever papal conference call to space. The day ended with Fincke and Feustel preparing their EMUs for the second spacewalk the following day.

"You are our representatives spearheading humanity's exploration of new spaces and possibilities for our future."

Extract from Pope Benedict's message
to the crew on board the ISS and Endeavour

"Holy Father, I felt your prayers and everyone's prayers arriving up here where outside the world ... we have a vantage point to look at the Earth and we feel everything around us."

Paolo Nespoli, replying to the Pope's concerns about how the Italian astronaut was holding up following his mother's death while Nespoli was in space. When Nespoli ended his five-month Expedition 26/27 residency aboard ISS, he brought back a silver medal that fellow ESA astronaut Roberto Vittori had taken to space. The medal, which depicted Michelangelo's 'Creation of Man' (a painting on the Sistine Chapel's ceiling) had been given to Vittori by the Pope.

The highlight of Flight Day 7 was the sixth longest spacewalk (and second longest during the ISS program) in history, a marathon 8-hour 7-minute affair that was more than 90 minutes longer than planned. Conducted by Feustel and Fincke, the EVA featured various maintenance and housekeeping tasks that included lubricating rotary joints, installing stowage beams on the trusses and attending to the upkeep of Dextre. Flight Day 8 was devoted to PR activities that included a chat between Vittori and Italian President Giorgio Napolitano. At the end of the day, the STS-134 crew (Table 4.13) headed for their sleeping quarters while the Expedition 27 crew boarded the Soyuz and headed for home, leaving Expedition 28 crewmembers Andrei Borisenko, Aleksandr Samokutyayev and Ron Garan on board the station. Flight Day 9 continued the previous day's PR theme, as Johnson and Chamitoff conducted a series of interviews with media outlets across the United States. This was followed by more transfer and organizational tasks, performed by Johnson and Vittori, and the pre-EVA preparations that were completed by Feustel, Fincke and Chamitoff. Flight Day 10 kicked off with the third spacewalk, the first to make use of a new pre-breathe protocol termed In-Suit Light Exercise (ISLE – this did away with the standard camp-out procedure and instead had the astronauts breathe oxygen for 60 minutes in the airlock at a pressure of 10.2 psi). Once Feustel and Fincke exited the airlock, they got to work installing the Power Data Grapple Fixture, after which they routed new power cables from Unity to Zarya before rounding out the EVA by installing a wireless video system. Meanwhile, inside the station, stowing activities continued as Johnson, Vittori and Garan moved new equipment and supplies into the ISS.

Flight Day 11 was highlighted by another inspection of Endeavour's TPS and preparation for the fourth and final spacewalk, which would be conducted by Fincke and Chamitoff

Table 4.13 STS-134

Mission	
Mission type:	ISS logistics
Mission duration:	15 days, 17 hours, 38 minutes, 22 seconds
Distance Travelled:	10,477,185 kilometers
Orbits completed:	248
Spacecraft	
Spacecraft:	Shuttle *Endeavour*
Launch mass:	121,830 kilograms
Landing mass:	92,240 kilograms
Payload mass:	15,770 kilograms
Crew	
Crew size:	6
Members:	Mark E. Kelly
	Gregory H. Johnson
	E. Michael Fincke
	Roberto Vittori (ESA)
	Andrew J. Feustel
	Gregory E. Chamitoff
EVAs	
Total EVAs:	4
Total duration:	28 hours, 33 minutes
Launch	
Launch date:	May 16, 2011, 12:56:28 UTC
Launch site:	Kennedy LC-39A
Landing	
Landing date:	June 1, 2011, 06:34:50 UTC
Landing site:	Kennedy SLF Runway 15
Docking with ISS	
Docking port:	PMA-2 (Harmony forward)
Docking date:	May 18, 2011, 10:14 UTC
Undocking date:	May 30, 2011, 03:55 UTC
Time docked:	11 days, 17 hours, 41 minutes
Orbital parameters	
Perigee:	321 kilometers
Apogee:	343 kilometers
Inclination:	51.6 degrees
Period:	91.17 min

the following day (this time, the spacewalkers did away with the ISLE option, since it had been found to use excessive carbon dioxide scrubbing capability). Flight Day 12 and EVA 4 began with Fincke and Chamitoff installing the Orbiter Boom Sensor System on the S1 Truss, after which the duo released some torque on the bolts holding the spare arm for Dextre. This final EVA of the Shuttle program saw the cumulative total time spent conducting EVAs at the station pass the 1000-hour mark since assembly began in December 1998. Another milestone belonged to Fincke who, having spent more than 377 days in space, surpassed the previous U.S. record held by Peggy Whitson. Flight Day 13 was dedicated mainly to maintenance and stowing tasks, in preparation for departure the following day. On Flight Day 14, after final farewells, the hatches closed between *Endeavour* and the ISS and it was time to head home. Undocking occurred on Flight Day 15 and Johnson backed the Shuttle away before performing the standard loop around the station. Re-entry and landing occurred on Flight Day 17 as *Endeavour* touched down at KSC, completing its 25th and final mission.

RICHARD GARRIOTT

Richard Garriott, or Richard Garriott de Cayeux to give him his full name (he changed his name when he married Laetitia de Cayeux in 2011), is also known by his alter egos 'Lord British' (in the video game *Ultima*) and 'General British' (in the video game *Tabula Rasa*). Born in Cambridge, England, Garriott was raised in the United States by his American parents, Helen and Owen Garriott. His father was also one of NASA's astronauts, who flew on Skylab 3 in 1973 and STS-9 ten years later. The catalyst for Richard Garriott's flight into space was his aptitude for writing computer games, allied with a keen business sense

Figure 4.30:   Richard Garriott. Credit: Russian Space Agency

Figure 4.31: Soyuz TMA-13 mission patch. Credit: Russian Space Agency

that led him to creating *Ultima I* and establishing his own publisher (Origin Systems), which he sold for $30 million in 1992. With his windfall, Garriott invested in Space Adventures and bought a ticket to become the first spaceflight participant. But when the dot-com bubble burst in 2001, Garriott had to sell his ticket to Dennis Tito. Frustrated but far from discouraged, Garriott knuckled down to creating more games to make more money. Which he did.

When he had made enough, he paid a non-refundable deposit to Space Adventures for his ticket to space. But his problems weren't over. Garriott's next challenge came in the form of a hemangioma on his liver. For most people this would not be a problem, but for an astronaut, a hemangioma[11] could be a death sentence in the event of a rapid decompression because of the risk of potentially fatal bleeding. Garriott was given two options: give up his deposit or risk surgery. Being a habitual risk-taker (he had participated in deep-sea submersible expeditions, sky-dived, visited the Antarctic, and explored jungles), Garriott opted for the surgeon's knife and today sports a 16-inch scar which he shows to audiences when giving lectures around the world. With the medical issue out of the way, Garriott embarked on twelve months of mission training and on October 12, 2008, he accompanied Expedition 18 crewmembers Mike Fincke and Yuri Lonchakov on board Soyuz TMA-13 (Figure 4.31, Figure 4.32) for his flight to the ISS (Garriott's father was present for the launch and landing). He returned on October 24, 2008, with Expedition 17 cosmonauts Sergei Volkov and Oleg Kononenko on board Soyuz TMA-12.

---

[11] At the time of this medical, Garriott also had several other medical conditions that would have disqualified him from serving as a career astronaut. These conditions included previous bilateral photorefractive keratectomy for myopia and a cross-fused ectopic kidney.

Figure 4.32: Richard Garriott, Yuri Lonchakov and Mike Fincke. Credit: Russian Space Agency

5

It was a beautiful launch

Figure 5.0. Credit: ESA

© Springer International Publishing AG 2017
E. Seedhouse, *Tim Peake and Britain's Road To Space*, Springer Praxis Books,
DOI 10.1007/978-3-319-57907-8_5

"For decades, mankind has dreamt of space travel and the final frontier, and from today the UK will trigger the next scientific and innovation revolution to turn science fiction into science fact. Not only are we celebrating the launch of the first UK Government-backed astronaut, but our first ever space policy will build on the inspiration he provides to grow our burgeoning space industry and bring space back down to Earth. Historically we haven't been a major player in space programmes; this policy will change that because, in the words of my hero Mr. Spock, to do anything else would be highly illogical."

> *UK Secretary of State for Business Innovation and Skills, Sajid Javid,*
> *December 2015, less than one week before Peake's launch.*

LAUNCH SITE

It is a bleak, windswept and almost featureless landscape that looks more like an artillery test range than a place to launch rockets. The few scraggly trees that manage to survive here have their lower trunks painted white to prevent the bark from cracking during what is always a brutal winter freeze. And in the middle of this dust-blown, camel-tracked panorama sits the lonely town of Baikonur (Figure 5.1). It is a typical Soviet-era block-style

Figure 5.1. Baikonur City. Credit: NASA/Bill Ingalls

apartment town: not a place that anyone punches into Expedia to book a tourist trip to (try doing that and you'll get the following message: *We were unable to find what you're looking for. Try searching by city name*: That's Expedia-speak for 'what the hell are you thinking?'), although the town is reachable by flying to Kyzylorda (prepare for a 24-hour-plus trip from the east coast of the U.S.). But if you happen to be an astronaut, Baikonur has more appeal than Disneyland has to the average five-year-old.

When the Soviets decided to get into the business of launching humans into space, they chose this barren wasteland for two reasons; its remoteness and the nearby Syr Darya River. The name Baikonur actually means 'rich soil'. That's an appropriate appellation for the original Baikonur, which was built hundreds of kilometers north of the current version. Baikonur Mk. 1.0 was a real town, next to which the Soviets built a fake launch facility to confuse any Western spies searching for such launch facilities during the Cold War. Of course, once the Soviets started launching rockets from the *real* launch site, it was only a matter of time before American spy planes put two and two together. Their cover blown, the Soviets decided to name the real launch site Baikonur as well. Since 2011, following the Shuttle's retirement, the site has witnessed all Russian, European and American manned spaceflight launches; all from the same pad that sent Yuri Gagarin into the history books.

If you decide to book a trip to Baikonur, you will probably land at Krainiy airport. From there, you will take a taxi, and the first indication that you have arrived in Baikonur will be the fisherman monument, a nod to the size of the fish that swim in the nearby Syr Darya River. If you happen to be visiting a friend, you continue driving to the main residential area on the banks of the river, near the Tyuratam railroad station (this is where you board the train if Moscow or Tashkent is your destination). The town has had many names since its founding in 1955. Originally it was known as Site 10, before being renamed Desyataya Ploshadka, Leninskiy and then Zvezdograd. Finally, in the 1990s, President Boris Yeltsin decreed that the town should be called Baikonur and the name stuck. As with so many Soviet towns, Baikonur began life as a military headquarters, before adding a cement factory and a rocket propellant plant to its facilities.

As the cosmodrome was being developed, soldiers and junior officers 'lived' in tents, while officers with families were lucky enough to find sturdier accommodation in nearby Kazakh villages. It was a grim existence, made worse by the lack of a water supply line. That didn't arrive until 1957, the same year that the air-strip near Tyuratam was completed. But as the Soviet lunar program began gathering momentum in the 1960s, Site 10 was transformed into what can only be described as an urban sprawl. By the 1970s, Site 10 boasted numerous stores, 356 apartment blocks, several cinemas, nine schools, 18 hotels, a sports stadium, health-care facilities ... and two more cement factories. The town grew so big that consideration was given to constructing a nuclear power station nearby to address the energy shortfall. Things were looking up ... until the crisis that came following the 1991 collapse of the Soviet Union. In a matter of months, Baikonur transitioned from a town on the up and up to one marked by crumbling infrastructure, a balky water supply, and vandalism. Two years later, the bleak climate and financial austerity had conspired to wreak more havoc on the town, with hepatitis-tainted drinking water and residents desperate to leave. In the winter of 1993, Baikonur was readying a new *Mir* mission, an occasion that drew a group of Russian officials and media, who got an unexpected earful of resident protests and colorful accounts of the hopelessness of their lives.

Figure 5.2. Hotel Sputnik. If you are thinking of taking a trip to Baikonur and are looking for a place to stay, here are a couple of reviews from TripAdvisor:

"Perfectly good enough stopping place on your way to the launch pad" (Reviewed March 18, 2016). Reasonably modern, reasonably well-furnished, clean, quiet – all the things you need to sleep well and soundly. This said, it's not the kind of place you'd plan a romantic weekend to. Location is at the edge of town, but you'll be unlikely to go for a stroll anyways. Food is on the heavy side, hardly what I call 'healthy fare'.

"Convenient hotel for visits to watch launches" (Reviewed September 5, 2015). This hotel is a nice respite in the Kazakh steppe when you're going to Baikonur to watch a rocket launch. The staff is friendly and helpful, even if they can't easily communicate with you. The breakfast is vast and delicious. The disco pub has normal western prices for drinks, and if you're nice to the bartender, they might let you take your drink up the stairs behind the bar to the roof where you can get a nice view of the night sky. *Room Tip:* All rooms have blackout curtains so you can sleep at any time – important when you're adjusting your sleep schedule to a rocket launch."

In 1995, conditions improved a little when a new Russian administration was formed and the Russian federal budget earmarked 761 billion rubles for Baikonur. Unfortunately, only 296 billion rubles was put on the table, which meant that upgrades to the city's infrastructure were curtailed. But gradually, more money was allocated to Baikonur and some semblance of normality returned. Five years later, the city featured a restaurant with Finnish vodka, Internet access, a local television channel, and new hotel (Figure 5.2) called the Sputnik ("decent, but terribly overpriced," according to TripAdvisor) which, with its $250-per-night rate, is mostly reserved for guests of the spaceport.

In the 1950s, Soviet scientists developing the R-7 and Burya ICBMs began searching for a new test site that incorporated an unobstructed view for hundreds of kilometers, a requirement dictated by the means of guiding these missiles to their targets. So, under a decree to search for a test site for long-range ICBMs, General Vasili Voznuk, the chief of Kapustin Yar (the current site at the time), led a search commission to find the new location. That location turned out to be Tyuratam. It was sparsely populated, linked to the industrial centers of the Soviet Union, and it even had an open pit mine (originally excavated by

Gulag prisoners) which could conceivably serve as a flame trench. But that was where the good news ended, because the climate was … well, it wasn't somewhere you'd choose to spend your vacation. The climate featured regular dust storms, summer temperatures that hit 50°C and winters that plunged the mercury down to negative double digits, and as if that wasn't bad enough, the region was a plague risk. With no infrastructure to speak of, Tyuratam sat firmly in the 'start-up' category of launch facilities. In fact, Sergei Korolev, the chief rocket designer at the time, had placed Tyuratam at the *bottom* of his list of choices. That didn't stop General Voznuk from deploying a surveillance team to the site, however, and based on the feedback from the team, Voznuk suggested the remote and inhospitable site be chosen. And so, in 1955[1], an engineering and construction brigade, an airfield construction division, a mining construction brigade, three automobile battalions and the brigade responsible for constructing the launch complex itself, rolled onto the Kazakh steppe to begin work on what would become one of the most famous launch sites in history.

What they found was … well, nothing much at all, because Tyuratam – which started its life as a settlement of cattle farmers – comprised little more than a few dozen residents living in no more than 30 houses. From a security perspective, this was fine, because the Soviets were determined to keep the project top secret; so secret, in fact, that even the most senior managers involved in the project were kept in the dark about what they were actually building. But even with the shroud of secrecy, it wasn't long before the Americans captured the first aerial photos of the test range, thanks to U-2 over-flights in 1957. A more high-profile U-2 flight occurred in May 1960, when Gary Powers took off from an American base in Pakistan, prompting Tyuratam air-defense crews to activate their guidance radar. Too early, as it turned out, because Powers was alerted to the danger and bypassed the site. After Powers was shot down, the U.S. decided to rely more on space-based surveillance and there was little the Soviets could do to disguise their activities at Tyuratam.

On April 12, 1961, almost a year after Powers was shot down, Yuri Gagarin sat atop his Vostok spacecraft on Tyuratam's Pad 1, waiting to kick-start the manned spaceflight era. But the success of Gagarin's flight created a headache for the Soviet leadership in the Kremlin, because for the flight to be ratified as a first, the Soviet Union had to disclose its launch site. That was out of the question, so the Soviets continued to deny the existence of Tyuratam, insisting instead that Gagarin's flight had launched from a cosmodrome near Baikonur. So Baiknonur, some 250 kilometers northeast of Tyuratam, became a cover for the Soviet Union's most famous spaceflight for no other reason than it just happened to be the first identifiable location downrange from Tyuratam. At the time, Baikonur was known only as a place of exile, where Gulag prisoners had worked in the copper and coal mines until the death of Stalin in 1953.

Eventually, the Soviets gave up the senseless deception and the awkward dual identity was put to rest, with the Soviets releasing the name 'Cosmodrome Baikonur' to the press. That veil of secrecy was lifted a little higher with the visit of French President Charles De

---

[1] The new test range was approved on June 2, 1955, by the Chief of Staff of the Soviet Ministry of Defense. This date became the official "birthday" of the Baikonur Cosmodrome.

Figure 5.3. Baikonur Cosmodrome. "Gagarin's Start" launch pad, October 2008. Public domain. Spaceflight is a superstitious arena, which should come as no surprise given that space-farers begin their journey to orbit by sitting on top of 300 tonnes of rocket fuel. Back in 1961, when Yuri Gagarin started the business of flying humans into space, he began the tradition of pre-launch rituals that persist to this day. Here are a few. Cosmonauts are expected to 1) Urinate on the right back wheel of the crew bus. 2) Plant a tree to honor those who have flown into space, as well as those who didn't make it back alive. 3) Watch *'The White Sun of the Desert'* movie, which is a Russian take on the Western and has been watched by every cosmonaut since 1970. 4) Have a haircut. 5) Have their rocket blessed by an Orthodox priest. 6) Drink champagne at breakfast on the day of launch. 7) Sign their hotel room door. 8) Be serenaded by a song called *The Green Grass Near My Home.* 9) Visit the Memorial Wall and lay red carnations. 10) Take a toy picked by their kids and dangle it from the Soyuz instrument panel.

Gaulle and East European officials in the 1970s, although the Soviets still went to some lengths to disguise the military uses of the cosmodrome. The military secrecy was sustained into the 1990s, with the Soviets referring to the site by its official and secret name of State Test Range No. 5 (or GIK-5). In fact, it wasn't until the latter half of the 1990s, when the civilian space agency Roskosmos took control, that the military presence began to be scaled down. Ultimately, GIK-5 was disbanded in April 2008, and eight months later the only military presence left was an air squadron in charge of R-36 missiles.

By that time, Russia and Kazakhstan had finally reached a rental agreement for Baikonur (Figure 5.3), with Russia agreeing to pay a fee of $115 million, although the

Table 5.1 Baikonur Facilities

- Oxygen and nitrogen plant for producing cryogenic propellants
- 3 fueling facilities
- A power station
- 600 energy-converting stations
- 92 communication sites
- 2 airports
- 470 kilometers of railways
- 1,281 kilometers of roads
- 6,610 kilometers of communication lines
- 360 kilometers of pipelines
- 1,240 kilometers of waterlines
- 430 kilometers of sewer lines

usage of the site was subject to administrative strife for a number of years afterwards because officials couldn't agree on exactly what constituted 'usage' (Table 5.1)[2].

A tourist's guide to Baikonur

The launch complex that is Baikonur can be divided into three distinct regions, each attached to a major player in Russian rocketry. The central – or Korolev – region, which is the area east and west of the original launch complex, is the site of the R-7 facilities and also the site where the R-9 ICBM was tested. The development of ICBMs eventually gave way to the Soviet manned lunar program in the 1970s and to the Energia-Buran program in the 1980s, each program being run by Korolev. The easternmost – Yangel – area was also originally used to develop ICBMs, most notably the R-16 which was developed by Mikhail Yangel. Like the Korolev area, this site eventually gave way to more conventional rocketry, when it witnessed testing of the Cosmos-1 launcher and the Zenit-2 rocket. On the western – Chelomei – side of the site, Vladimir Chemolei's OKB-52 design bureau busied itself developing successive generations of the UR-100 ICBM, before turning their attention to the Proton rocket.

Another feature that may grab your attention while you're wandering around are the railroad tracks and the trains that are used to transport the launchers to the site. Perhaps one of the most enduring icons of the launch complex, the *motovoz*, as the diesel-powered trains are known, have been in business for decades, shuttling between the town and the launch sites. In the 1950s and 1960s, passengers would suffocate in the train cars during the summer as the stifling climate turned the compartments into saunas; in the winter, the occupants would freeze when snowstorms descended onto the steppe, often trapping

---

[2] When a Proton rocket crashed in 2007, the launch site was shut down for two months while the Russians demanded compensation from the Kazakhstan government (which paid the Russians $61 million). Since then, the usage of the site – which is valued at $3.4 billion – has been subject to dozens of interim agreements, ranging from the launch of ICBMs to the use of dormant infrastructure. In 2015, with the Russians having decided to move their launch operations to an alternative – and less trouble-some – site, the Kazakhs started looking for other customers, including ESA and Israel.

workers for days at a time. The simple solution would have been to house the workers closer to the launch facilities, but this idea was a non-starter because of the highly-classified nature of the work that was being conducted at what was effectively a military test range.

Today, thanks to a road that follows the main rail track, tourists – or accidental travelers – can drive the route from the town of Leninsk to the cosmodrome, passing the village of Tyuratam along the way. The road is designated the M-32 highway but this isn't your typical highway, thanks to a section of no-man's land, the check-points, myriad antenna dishes, the propellant-producing plant, and the towers of the command bunkers in the distance. After driving 20 kilometers across the steppe you come across two junctions, one leading to the east where Soyuz and Zenit launch complexes can be found, and the other route that leads to Proton City.

LAUNCH VEHICLE

Major Peake's ride to the ISS was the Soyuz TMA-M (see Appendix III, Tables 5.2a through 5.2d, and Figure 5.4), a spacecraft that has a heritage stretching all the way back to 1966 when the first unmanned flight was launched. In the rush to catch the Americans in the race to the Moon, the system was prematurely declared man-rated, which was unfortunate for Vladimir Komarov, the lone occupant for the spacecraft's first manned flight, Soyuz 1, in 1967. After a series of technical problems on orbit, the main parachutes failed to open during the descent and Komarov was killed. Since Komarov's flight, there has been only one other failure of the vehicle, on Soyuz 11 in 1971, when a premature cabin depressurization killed the crew of three. Despite these four deaths, the Soyuz remains one of the safest manned crew vehicles in history (14 astronauts were killed in the two Shuttle accidents), especially when you consider its longevity. Over the years, it has been used to ferry cosmonauts to military space stations, to Salyut stations, and to *Mir*. In 2017, it is the only manned spaceflight vehicle currently available for launching crews to ISS.

Table 5.2a Soyuz TMA-M

Type:	Soyuz TMA-M
Manufacturer:	RKK Energia
Length:	7.48m
Diameter:	2.72m
Launch Mass:	7,150kg
Span:	10.6m
Modules:	3
Habitable Volume:	8.5m³

Table 5.2b Soyuz Orbital Module

Length:	2.98m
Diameter:	2.26m
Launch Mass:	1,300kg
Habitable Volume:	5m³
Attitude Control:	4 DPO Thrusters
Rendezvous:	KURS

Table 5.2c   Entry Module

Length:	2.24m
Diameter:	2.17m
Launch Mass:	2,950kg
Habitable Volume:	3.5m³
Thermal Protection:	Ablative Heat Shield
Attitude Control:	24 Thrusters
Propellant:	Hydrogen Peroxide
Landing:	Parachutes
Pilot Chute:	4.5m²
Drogue Chute:	16m²
Main Chute:	518m²
Landing Engines:	6
Landing Speed:	2-4m/s
Cargo Downmass:	100kg
Flight Computer:	KSO 20M

Table 5.2d   Instrumentation Module

Length:	2.26m
Diameter:	2.72m
Launch Mass:	2,900kg
Habitable Volume:	None
Main Engine:	SKD (Redundant)
Trust:	2,942N
Attitude Control:	28 DPO Thrusters
Thrust:	26.5N/130N
Oxidizer:	Nitrogen Tetroxide
Fuel:	Unsymmetrical Dimethylhydrazine
Propellant Mass:	800kg
Power Generation:	2 Solar Arrays
Span:	10.6m
Area:	10m²
Power:	1000W
Flight Computer:	TsVM-101

The Soyuz vehicle (see Appendix III) comprises three modules; the Orbital Module, the Instrumentation Module and the Entry Module. These are each equipped with separation mechanisms, to enable separation following the deorbit burn. Certified to remain docked with the ISS for six months, the Soyuz not only serves as a means of ferrying crews to and from the orbiting outpost, but also as a lifeboat in case of emergency.

The uppermost module is the Orbital Module which, as its name suggests, is the module the crew lives in during their trip to and from the station. With only five cubic meters of habitable space (the Gemini capsule had barely more than two cubic meters, but that was for just two crewmembers), the confines of the Soyuz are snug at best and this is not a spacecraft for the claustrophobic! In addition to the KURS hardware system that is used for rendezvous and docking procedures, this module also features various electrical and communications interfaces for use when the Soyuz is attached to the station. The life support system can support a three-man crew for up to 30 days.

Figure 5.4.   TMA-07M docked with the Rassvet module, 2012. NASA. Credit: NASA

The middle module is the Entry Module, which features another 3.5 cubic meters of habitable area and also houses the control systems and crew seats. During launch and landing, this is the module in which the crew can be found, checking system performance on the vehicle's primary KS0 20M computer. This module also features 24 hydrogen peroxide thrusters for maneuvering, a parachute system, and a cargo compartment that can carry 100 kilograms of essentials.

The Instrumentation Module, also known as the Service Module, is the lowermost module. As its name suggests, this is the module that houses all the support equipment. Everything from thermal control systems to telemetry, and electrical power systems to navigation equipment can be found here. This module also features the vehicle's liquid-fueled propulsion system that is capable of performing attitude control maneuvers, orbit adjustments, and approach and rendezvous burns. The propulsion system comprises 28 thrusters, 12 of which provide low thrust (26.5 Newtons) for precise maneuvering, while the other 16 provide a bigger punch (130 Newtons) for more sustained maneuvers.

Another element of the Soyuz vehicle is the launch abort system, which provides the crew with launch and abort capabilities throughout the ascent phase. Perched on top of the Soyuz rocket is a tower that provides a means of escape during the first two-and-a-half minutes of flight. If things go awry after that, the mode of abort is separation of the modules, but let's deal with the tower first. Termed the Launch Escape Tower (LET), this feature can be triggered either automatically or by the crew. If something goes wrong on the launch pad, the crew can be carried to an altitude of 1,500 meters and up to four kilometers downrange. If a contingency event occurs shortly after launch – such as decompression, loss of pressure in the combustion engines, or loss of control, for example – the abort command is initiated and the abort motors mounted on the LET do their job, carrying the spacecraft and payload fairing away from the launch vehicle. Burning for six seconds, the abort motors provide a punchy 723 Kilonewtons of thrust, which is more than enough to carry the crew to a safe distance to deploy the parachutes and execute a nominal landing.

The booster

The Soyuz launcher has been around for a long while now. In fact, it has a history that stretches all the way back to 1957. Based on the R-7 missile, which made almost 30 launches between 1957 and 1961, the Soyuz Launch Vehicle has undergone myriad transformations and reconfigurations over the years and has been used for just about everything, from ferrying crews and launching cargo to orbit, to hauling satellites into space – you name it, the Soyuz has done it. The current manned iteration, the Soyuz FG (see Figure 5.5 and Tables 5.3a through 5.3d), is a fairly slender vehicle, with a diameter of less than three meters (ten with the strap-on boosters), and stands less than 50 meters tall.

Figure 5.5.   Soyuz launch vehicle. Credit: Russian Space Agency

Table 5.3a Soyuz Launch Vehicle

Type:	Soyuz FG
Manufacturer:	TsSKB-Progress
Height:	49.5m
Diameter:	2.68m
Launch Mass:	305,000kg
Stages:	Boosters + 2 Stages
Boosters:	4
Mass to LEO:	7,100kg
Success Rate:	100%

Table 5.3b Boosters

Boosters:	4
Diameter:	2.68m
Length:	19.6m
Empty Mass:	3,810kg
Fuel:	Rocket Propellant 1
Oxidizer:	Liquid Oxygen
Fueled Mass:	43,410kg
Propulsion:	1 RD-107A
Propellant Feed:	Turbopump
Comb. Chambers:	4
Total Thrust SL:	838.5kN
Total Thrust Vac:	1,021kN
Engine Length:	2.58m
Engine Diameter:	1.85m
Engine Dry Weight:	1,090kg
Burn Time:	118s
Specific Impulse:	264s (SL) 310s (Vac)

Table 5.3c Core Stage

Diameter:	2.95m
Length:	27.8m
Empty Mass:	6,550kg
Fuel:	Kerosene
Oxidizer:	Liquid Oxygen
Fueled Mass:	99,500kg
Guidance:	From 3rd Stage
Propulsion:	1 RD-108A
Engine Type:	Gas Generator
Propellant Feed:	Turbopump
Total Thrust SL:	792kN
Total Thrust Vac:	990kN
Engine Length:	2.87m
Engine Diameter:	1.95m
Engine Dry Weight:	1,075kg
Burn Time:	280s
Specific Impulse:	258s (SL) 321s (Vac)
Throttle Capability:	Yes
Restart Capability:	No

Table 5.3d Block I Upper Stage

Diameter:	2.66m
Length:	6.74m
Empty Mass:	2,410kg
Fuel:	Rocket Propellant 1
Oxidizer:	Liquid Oxygen
Fueled Mass:	25,300kg
Propulsion:	1 RD-0110
Engine Length:	1.58m
Engine Diameter:	2.24m
Engine Dry Weight:	408kg
Specific Impulse:	326s (Vac)
Restart Capability:	No
Oxygen to Fuel Ratio:	2.2

As you can see in Figure 5.5, the Soyuz FG comes equipped with four strap-on boosters that provide the extra oomph during the first phase of the ascent. The liquid-filled boosters ignite prior to lift-off and are discarded once the fuel has been exhausted. The tapered part of the booster is the oxidizer tank, while the cylindrical portion houses the fuel tank. Measuring almost 20 meters in length and almost three meters in diameter, each booster is fitted with an RD-107A engine that provides 838 Kilonewtons of thrust. Unlike the Shuttle Main Engines, the Soyuz engines can't be gimballed, although the system is fitted with Vernier jets that may be used for attitude control. The easiest way to work our way around the launcher is to take it stage by stage. We'll begin with the core stage.

The core, which serves as the first and second stage, powers the launcher after the boosters are jettisoned. Simplistic in design, the stage comprises an oxidizer and fuel tank that holds nearly 100,000 kilograms of propellant. The interface between this stage and the boosters is a load-carrying ring that transfers loads from the boosters to the vehicle. Perched on top of the core stage is a truss segment that latches on to the upper stage. The RD-108A engine that powers the core stage is similar to the booster main engine. Once the stage has completed its 290-second burn, it separates from the third stage thanks to the push created by ignition of the third stage. Like I said, a fairly simple – but extremely reliable and effective – system. Rounding out the launcher is the third stage, powered by a RD-0110 engine that features a single turbopump capable of providing nearly 300 Kilonewtons thrust for the 300-second burn time of the stage.

Launch Abort System – Russia's *Challenger* event, September 1983
As long as there have been manned rockets, astronauts have known that the most dangerous phases of any mission are the final minutes before, during and immediately after launch. Manned spaceflight's first launch pad contingency occurred in April 1975, when Soyuz 18A suffered a major glitch during the ascent which resulted in the capsule being sent on a suborbital trajectory that ended in the middle of Siberia. Eight years later, the Soyuz 18A flight may have been on the minds of Flight Engineer Gennadi Strekalov and Commander Vladimir Titov as they

clambered on board their Soyuz T-10. One person who definitely had a bad feeling about the flight was Strekalov's mother, Praskovya Strekalova. She had good reason to worry, because six months earlier her son had flown on board Soyuz T-8 to the Salyut 7 space station, only to discover that the docking mechanism was balky, forcing a return to Earth.

This time around, the cosmonaut's destination was the same, and Strekalov had reason to be a little more relaxed. After all, what were the chances of two malfunctions in the space of six months? Pretty good, as it turned out. You see, trouble is never far away when it comes to manned spaceflight, and sure enough, trouble reared its ugly head just 90 seconds before lift-off. A fuel valve failed to close, propellant spewed across the launch pad and caught fire, and a conflagration almost engulfed the rocket. Making matters worse, if that was possible, the separation pyrotechnics were armed, since the vehicle was just seconds from launch. A few kilometers away, watching in horror, was Flight Director Alexei Shumilin. Inside the cockpit, Titov and Strekalov knew something was very wrong, but they had no windows to look out of to gauge the severity of the fire.

The urgent chatter over the comm loop became increasingly distressed, and then downright panicky, when the controllers realized that the raging inferno had burned clean through the cables that carried the signals to fire the vehicle's escape rocket. As a result of some human factors design oversight, or just really, *really* bad design, the two cosmonauts had no way to activate the escape system manually. Their last chance of coming out of this alive was an override system, which required two controllers in different rooms more than 30 kilometers from the launch pad to press buttons within five seconds of one another. They could only do this if they received the code word *Dnestr* from Shumilin. Fortunately – somehow – that command was executed, and the escape system's pyrotechnics fired. The Soyuz Descent and Orbital Modules were wrenched away from the burning booster, with Titov and Strekalov subjected to more than 20 mind-numbing Gs.

From the vantage point of the blockhouse, all the witnesses could see was a red and yellow cloud, through which an object appeared as a rapidly-moving projectile. By this time, the capsule was bulleting along at Mach 1, rising nearly vertically through the holocaust of flames to an altitude of more than 1000 meters. Somehow, Titov and Strekalov had had the presence of mind to deactivate the cabin voice recorders, which was a good thing because what they were saying had no place on a transcript, according to Titov! After landing, copious amounts of vodka and cigarettes calmed frayed nerves. The mission had lasted five minutes and 13 seconds.

Once in low Earth orbit (LEO), the Soyuz deploys its solar arrays and communication antennae and begins on-orbit operations, which means chasing the ISS. Until 2011, this chase took the best part of two days, as the Soyuz performed a series of main engine burns to gradually raise its orbital altitude and make slight changes to its trajectory. As you can imagine, being a crewmember cooped up in the confines of the Soyuz for 48 hours was

Table 5.4 Soyuz TMA-19M mission

Mission	
Mission type:	ISS Expedition 46/47
Mission duration:	185 days, 22 hours, 11 minutes, 57 seconds
Distance Travelled:	10,477,185 kilometers
Orbits completed:	2,893
Spacecraft	
Launch Vehicle:	Soyuz FG
Crew	
Crew size:	3
Members:	Yuri I. Malenchenko (Roskosmos)
	Timothy L. Kopra (NASA)
	Timothy N. Peake (ESA)
Back-ups:	Anatoly A. Ivanishin (Roskosmos)
	Kathleen H. Rubins (NASA)
	Takuya Onishi (JAXA)
Launch	
Launch date:	December 15, 2015, 11:03:09 UTC
Launch site:	Baikonur Cosmodrome, Site 1, Pad 5
Landing	
Landing date:	June 18, 2016, 09:15:06 UTC
Landing site:	47° 24 N, 69° 42 E
Docking with ISS	
Docking port:	MIM1 Rassvet, -Y axis, Russian Segment
Docking date:	December 15, 2015, 17:33 UTC
Undocking date:	June 18, 2016, 05:52 UTC
Orbital parameters	
Perigee:	398 kilometers
Apogee:	410 kilometers
Inclination:	51.6 degrees
Period:	92.6 min

anything but comfortable. Fortunately, with some fine-tuning, more than 40 hours has been trimmed from the transition from LEO to the ISS, and the launch-to-docking time is now a shade over six hours. Now that we're familiar with the nuts-and-bolts operation of the Soyuz, let's return to Major Peake's mission.

Preparations for launch

Two weeks ahead of the target launch date of December 15 for Soyuz TMA-19M (Table 5.4), the prime and back-up crews performed a familiarization training session on the vehicle as it sat inside the processing facility at Site 254. While the crew was busy, mission managers gave the green light to load propellants and pressurized gas into the spacecraft, a procedure that was completed two days later. With the propellant loading complete, the spacecraft was returned to the processing facility in preparation for the next phase of pre-launch tasks. The following day, the spacecraft was locked onto the launch vehicle adapter and placed into the horizontal position, ready for an inspection prior to being covered with the payload fairing. Four days later, on December 8, engineers performed a final inspection of the spacecraft and the fairing was placed over the vehicle. The next day, the payload section was transported from Site 254 to Site 112, the vehicle processing building, where it was integrated with the Soyuz-FG. Two days before launch, on December 13, the whole kit and caboodle was rolled out to the launch pad at Site 1. Everything was set.

LAUNCH DAY

Major Tim Peake's final steps to becoming Britain's first male astronaut began with him signing his hotel room door, just one of many tributes to Yuri Gagarin. At 08:30, Peake was suited up and made his way out of Baikonur Cosmodrome Building 256, together with NASA's Tim Kopra and Russian cosmonaut Yuri Malenchenko. Waddling awkwardly in their bulky spacesuits, the crew marched to the center of the building's courtyard before halting in front of a Russian space official. In keeping with the traditions of Russian manned spaceflight, the official asked each man if he was ready to start the mission. After answering in the affirmative, the crew turned to give one final wave to family members before boarding the bus that would take them to Baikonur's famous launch pad.

In between the events leading up to the launch, the Twitterati kept themselves busy by arguing over who was the first British astronaut. Some opined that Peake was Britain's first official astronaut, which was true if *official* meant government-sponsored. But, the Twitterati argued, Britain's first *true* astronaut was a female – Helen Sharman – and by 'true' the Twitterati meant that she was the first non-dual-nationality British person to fly into space, meaning that Michael Foale, Piers Sellers, Nicholas Patrick, Gregory Johnson (NASA astronauts), Mark Shuttleworth and Richard Garriott (spaceflight participants) didn't count. For some, the argument and counterargument all got a bit confusing.

> "This isn't a one-off mission. We have a serious project in the European Space Agency to land on the moon, and that is part of an exploration of the solar system that will eventually take us to Mars."
>
> *Tim Peake's comment the day before his launch to the ISS.*

As media attention converged on Baikonur, Google got in on the act and promoted Peake's flight on their search bar. Then, at 09:15, the subject of a British astronaut being launched to the ISS – #BritInSpace – became the leading hashtag trending on Twitter. The European Space Agency (ESA), no slouch in the online media department, duly tweeted a photo of the crew ensconced inside their spacecraft, together with a tagline explaining that they would be sitting there for almost four hours. Not that different from a London commute, in other words!

> "It's great. I can't wait to see this. This has been a long wait. It is just fab to get to this point."
>
> *Rebecca Peake, on saying goodbye to her husband for six months*

As the minutes ticked by, media outlets scrambled to fill in that part of the launch process in which nothing much happens by wheeling out various doom-and-gloom experts, whose job it is to warn viewers about all the potential disasters that can happen if everything goes sideways. At 10:00, David Cameron (British Prime Minister at the time) was supposed to have been saying his piece on the whole 'Brit in Space' event, but instead the only information posted on the Twitter feeds was a note about the value of the mission to the global space sector. Dry stuff. Thanks PM. Ten minutes later, the one-hour countdown was well underway, with ESA confirming that all systems were go. This announcement was followed by explanations on Twitter that Peake would be calming his nerves by listening to his choice of songs, which included Queen's *Don't Stop Me Now*, U2's *Beautiful Day*, and Coldplay's *A Sky Full of Stars*. At 10:25 the service structure was removed, and a few minutes later useful information about the distance from Earth to the ISS popped up on Twitter (it's about 225 miles which, for comparison if you happen to be British, is the distance between London and Scarborough). At 10:50, ESA updated its Twitter feed with an announcement that fueling was still ongoing. Launch was only minutes away (Figure 5.6).

11:03: Lift-off!

Launch took place as planned on December 15 (Figure 5.7), at 14:03:09.328 Moscow Time (11:03:09.328 GMT). Powered by four R-107A engines, the stack arced upwards and eastwards, aligning to the ascent trajectory it would follow to space. The vehicle's emergency escape system was jettisoned less than two minutes into the flight, followed by separation of the four boosters. The second stage took over, firing its booster for the next four minutes before handing propulsion duties over to the third stage. At 11:11:58 GMT, the spacecraft separated from the third stage, entering its initial parking orbit shortly afterwards (Table 5.5).

> "Tim, I know you have been dreaming of this day for a long time and we will be with you for every step of the way, watching with admiration and wonder. So, on behalf of everyone in Britain, let me wish you the very best of luck. You are doing us all proud."
>
> *David Cameron, British Prime Minister.*

Shortly after David Cameron's message, the first images of Peake entering LEO were beamed back to Earth. At 11:30, the Soyuz was traveling at an altitude of 210 kilometers and zipping around the Earth at 28,000 kilometers per hour. Ahead of the crew was a six-hour, four-orbit journey to catch up to the ISS, with a projected arrival time of around 17:00 GMT.

> "We would say this is once in a lifetime, Tim hopes it won't be. You know we'll wait and see on that. All the support from back home has been absolutely magnificent. I just think it's so wonderful that you've got all this international cooperation working to one end; medical and scientific experiments that are trying to benefit mankind, and they all work together happily and depend on each other, live together. It doesn't matter what nationality anyone is. And the wonderful back-up system right across the globe. It's that over-used word, awesome."
>
> *Nigel Peake, speaking to The Telegraph's Roland Oliphant in Baikonur*

At 12:50, ESA confirmed its first contact with its astronaut, an announcement that was followed ten minutes later by a message from NASA astronaut Scott Kelly, who said that he had spotted the Soyuz from the ISS. Shortly afterwards, Peake got to work on one of the many experiments he had been tasked with during his flight – the Space Headaches experiment.

Figure 5.6. Tim Peake, Tim Kopra and Yuri Malenchenko. Credit: AFP Photo Pool/Shamil Zhumatov

Figure 5.7. Launch. Credit: NASA

Table 5.5 Soyuz Parking Orbit

Parameter	Planned orbit	Actual orbit
Orbital period	88.64 minutes (+/-0.367)	88.74 minutes
Inclination	51.67 degrees (+/-0.058)	51.64 degrees
Perigee	200 kilometers (+7/-22)	200.75 kilometers
Apogee	242 kilometers (+/-42)	253.08 kilometers

One of the myriad occupational hazards associated with being an astronaut is the problem of migraines, caused by the translocation of fluids from the lower extremities to the upper extremities. Imagine another one-and-a-half to two liters of extra fluid in your chest and head area and you have some idea of what astronauts have to put up with. The Space Headaches experiment was designed to help Earth-bound researchers develop ways to alleviate the symptoms and hopefully improve astronaut performance on orbit, which is why Peake spent time en route to ISS checking answers on the two-page headaches questionnaire. At 13:45, the Soyuz executed its final burn to put the vehicle on course for the ISS. Meanwhile, more than 100 miles below, the normally staid House of Commons in London had been hit by Tim Peake fever, as ministers sang along to the tune of David Bowie's 'Space Oddity' while simultaneously unveiling the Government's new space policy – which included familiar phrases borrowed from a few well-known sources. Here's a taste:

"Major Tim Peake has successfully blasted into orbit. This morning, the Government launched its space policy, which achieved lift off. Launched a short time ago in a museum that is not far, far away, this policy document shows that there are no limits

Figure 5.8.   Soyuz docked with the ISS. Credit: NASA

to the UK's ambitions in this area. To mix intergalactic metaphors – we want to boldly go to infinity and beyond and our new policy will make it so."

That soundbite, from Business Secretary Sajid Javid, set the standard for other MPs to follow – and few of them resisted the temptation – but the essence of the message was the encouraging fact that the value of the UK space sector had doubled in the seven years preceding Peake's launch. By the end of 2015, space was worth some £11.8 billion and employed 37,000 people (and the new policy aimed to more than triple the value of the sector to £40 billion by 2030). A far, *far* cry from the Thatcher years. While the MPs got on with the job of congratulating themselves and patting each other on the back for resurrecting the British space industry, PM David Cameron was following the events of Peake's mission at home. Around the country, Tim Peake fever also extended to millions of schoolchildren, who had abandoned lessons to follow the mission.

Another mission update was tweeted at 15:45, as ESA announced that the crew were feeling well and would be arriving at the ISS at 17:24. This message was followed just 25 minutes later by ESA's announcement that the Soyuz was just 80 kilometers away from the ISS and that mission commander Yuri Malenchenko had taken over manual control. Another 25 minutes later, and the 80-kilometer distance had shrunk to just one.

At this point, it's worth taking a time out and explaining the complexities of orbital mechanics, because this rendezvous and docking procedure (see Appendix III, Figure 5.8, and sidebar) is anything but easy and is nothing like you see it portrayed in your average sci-fi movie.

Rendezvous and Docking: A Primer

Rendezvous and docking is all about two objects: an active vehicle – the *chaser* – and a passive vehicle – the *target*. To understand how these two objects rendezvous in space, we need to wrap our minds around the relationship between spacecraft velocity and orbit. A spacecraft that is placed in a specific orbit has a specific velocity, dictated by the altitude of that orbit. Spacecraft in lower orbits have faster orbital velocities than spacecraft in higher orbits, for the very simple reason that the force of gravity becomes weaker the further you are from Earth (angular velocity is part of the equation but we're trying to keep it simple here). For the chaser to rendezvous with the target, both must be in the same orbital plane and the phase of the orbit (the position of the chaser in that orbit) must match. To reach this point, the chaser is placed in a lower (faster) orbit than the target, which means that the chaser is traveling faster than the target. Then, by performing a series of carefully calculated burns, the chaser gradually boosts its orbit to the higher one where the target is. Once on the same orbital plane, the chaser performs some more finely tuned burns to gradually close in on the target, until it reaches the zone of proximity operations which is where everything slows right down. The actual process of docking takes place at a snail's pace of just three centimeters per second, but before we get to that stage there is the approach method to consider.

The methods of approach for conducting proximity operations depend on several factors, such as safety, the type of spacecraft and the mission timeline. In the *V-bar approach*, the flight path of the chaser follows the same direction as the orbital motion of the target in the horizontal plane. Another way of thinking about this is to imagine the motion of the chaser parallel to the target's velocity. A chaser that conducts a V-bar approach does so by firing small thrusters to gradually increase its velocity in the direction of the target. The effect of these small thruster firings is to gradually raise the chaser's orbit until it is in very close proximity with the target. The V-bar itself is a velocity vector that extends along a line directly ahead of the target – all the chaser has to do is to approach the target along this V-bar. Alternatively, the chaser may use the *R-bar approach*, which requires the vehicle to move along its radial vector either above or below the target. As with the Z-bar approach, the chaser fires its thrusters to adjust its altitude but the orbital velocity of the chaser does not change because the thruster firings are in the radial direction.

To reach the proximity operations phase, the Soyuz had performed a series of precisely calibrated pre-programmed maneuvers during its first, second and third orbits (Table 5.6) to raise its orbital altitude from 220 kilometers to 420 kilometers. This had brought the spacecraft to a point 300 meters from the ISS. With a Soyuz in its near phase, the commander usually hands over docking to the KURS automated docking system, but in this mission

Table 5.6 Soyuz Pre-Programmed Maneuvers

Orbit No.	Time	Firing duration	delta V	Period	Inclination	Resulting perigee	Resulting apogee
1	14:48:15	71.1 seconds	28.54 m/s	89.62 minutes	51.64 degrees	229.63 kilometers	297.29 kilometers
2	15:32:29	58.4 seconds	23.57 m/s	90.44 minutes	51.64 degrees	296.53 kilometers	323.45 kilometers
2	16:06:44	18.2 seconds	7.00 m/s	90.69 minutes	51.67 degrees	298.06 kilometers	336.85 kilometers
2	16:36:58	18.2 seconds	7.00 m/s	90.94 minutes	51.64 degrees	318.87 kilometers	337.34 kilometers

Table 5.7 Orbital Data

Launch Date: December 15, 2015
T-0 Time: 11:03:09 UTC
Docking: December 15, 2015
Time: 17:24 UTC
Docking Port: Rassvet

Soyuz Insertion Orbit

Perigee: 200 Kilometers (+7 / -22 Kilometers)
Apogee: 242 Kilometers (+/-42 Kilometers)
Inclination: 51.66 Degrees (+/-0.058°)
Period: 88.64 Minutes (+/-0.367min)
Phase Angle (to ISS): ~32 Degrees

Space Station Orbit

Orbit: 397 by 408 Kilometers
Inclination: 51.64 Degrees
Period: 92.62 min

the KURS[3] aborted the approach, forcing Commander Yuri Malenchenko to switch to manual control and guide the spacecraft through the rendezvous and docking process (Table 5.7). This led to a delay, but with the alignment confirmed the final approach was executed, until initial contact and capture were confirmed at 17:33 GMT (Table 5.8).

---

[3] Although Mission Control was unable to explain why the KURS had aborted the approach, an explanation was forthcoming thanks to a reporter working for *Novosti Kosmonavtiki* magazine. Apparently, the cause of the abort was a malfunction of the DPO-B No. 20 attitude control thruster, which is used to make very precise attitude adjustments to the vehicle. Scrutiny of the video and audio feed revealed an alarm had sounded in the Soyuz cockpit, followed by an audible warning that a failure of the K1B manifold had been detected. Normally, such a failure – later diagnosed as a failed pressure chamber sensor – would prompt a switch to a backup circuit, but this didn't happen. Given that the incident echoed what had happened during the near catastrophic collision between a Progress cargo vehicle and the *Mir* station in 1997, the malfunction generated much discussion.

Table 5.8 Rendezvous Profile

Launch	11:03:09
Orbital Insertion	11:11:58
Solar Array & KURS Antenna Deploy	11:12:06
Ground Station Loss of Signal	11:19
Rendezvous Burn 1 (dV: 28.54m/s; dt: 71.1s)	11:48:15
Rendezvous Burn 2 (dV: 23.57m/s; dt: 58.4s)	12:32:29
Ground Station Acquisition of Signal	12:29
End of Ground Station Pass	12:57
Rendezvous Burn 3 (dV: 7.00m/s; dt: 18.2s)	13:06:44
Target Orbit: 298.06 by 336.85 km, 90.69 min	13:06:44
Rendezvous Burn 4 (dV: 7.00m/s; dt: 18.2s)	13:36:58
Target Orbit: 318.87 by 337.34 km, 90.94 min	13:36:58
Ground Station Acquisition of Signal	14:05
End of Ground Station Pass	14:24
Automated Rendezvous Initiation	15:15:55
Rendezvous Burn 5	15:24
Trajectory Correction	15:46
VHF-2 Voice Link Established (Range: 200km)	15:48
KURS-A System Activation	15:49
KURS-P System Activation	15:51
Rendezvous Burn 6	16:10
KURS-P Validation, Range: 80km	16:14
Range: 33km; Range Rate: 45m/s	16:27
Range: 20km; Range Rate: 31m/s	16:32
KURS A&P Test, Range: 15km	16:34
Range: 13km; Range Rate: 12m/s	16:37
ISS Inhibits: RapidScat & SCAN Testbed	16:40
TV System Activation, Range: 8km	16:42
Range: 5km	16:47
Rendezvous Impulse	16:54
Rendezvous Impulse	16:58
Fly around Initiation	17:01:07
Station keeping	17:07:26
Final Approach	17:13:00
ISS Inertial Snap & Hold	17:15:30
Orbital Sunset	17:18
DOCKING - Free Drift	17:24:09
Hooks Closed - Safe Docking Mode	17:35
Leak Checks	17:50
ISS to USOS Control	18:20
Hatch Opening	19:25

Figure 5.9.   Peake is greeted by Scott Kelly on arrival at the ISS. Credit: ESA

"They say nothing in space flight is routine, and the plan on this mission has changed already. Apparently unhappy with the approach to the ISS plotted by the automated docking system, Commander Yuri Malenchenko took manual control of the Soyuz and backed the craft back from the station, opting to pilot the ship in himself. Chatting through the manoeuvre with ground controllers at Roscosmos' mission control centre in Korolyov, the headquarters of the Russian space program outside Moscow, Col. Malenchenko brought the Soyuz craft back to a position about 100 metres from the station until he could line up with the docking port. He then gently guided the craft back to the station, and finally docked with the ISS at 20:34 Baikonur time, somewhere over India."

> *Dispatch from* The Telegraph's *Roland Oliphant,*
> *from the scene at Baikonur*

"It was a beautiful launch. We just got on with the work, but that first sunrise was absolutely spectacular. I hope everyone had a great celebration. I hope you enjoyed the show. I think we had a great time in the office, that's for sure."

> *Major Peake's first dispatch from the ISS*

Finally, after leak checks had been completed, Peake and his colleagues entered the ISS, where NASA's Scott Kelly and Russian cosmonauts Mikhail Kornienko and Sergey Volkov were on hand to give them all a hug (Figure 5.9). Next on the 'to-do' list: on-orbit mission operations.

6

A Brilliant Ambassador

Figure 6.0.   Credit: ESA

© Springer International Publishing AG 2017
E. Seedhouse, *Tim Peake and Britain's Road To Space*, Springer Praxis Books,
DOI 10.1007/978-3-319-57907-8_6

TIM PEAKE'S FIRST THREE MONTHS ON ORBIT

When Major Peake entered the ISS, he did so as a member of the Expedition 46 crew (see Appendix IV), which ran between December 11, 2015 and March 1, 2016, and then transitioned to Expedition 47 (see Appendix IV), which started on March 1 and finished on June 5, 2016. The account of Peake's Expeditions is covered in this chapter and the next, but it is impossible to tell his story in isolation, so what follows is a snapshot of the crewmembers he shared his mission with. Also, since many readers may not be familiar with the complexity of operations on board the International Space Station (ISS), there are some brief descriptions of some of the routines that comprise day-to-day life on the orbiting outpost. But first, let's introduce Peake's crewmates.

Scott Kelly

Actually, Scott Kelly needs no introduction, because even those who have little interest in manned spaceflight will have heard about NASA's much-hyped one-year mission in space. Born February 21, 1964, in Orange, New Jersey, Kelly is a retired captain in the U.S. Navy and was a veteran of two Space Shuttle flights and one long-duration mission to the ISS before he blasted off to the ISS again, this time with mission crewmate Mikhail Kornienko. As with all astronauts, Kelly's list of qualifications would fill many chapters of a book, so what follows is a biographical snapshot of the events leading up to Kelly being selected as an astronaut.

After leaving Mountain High School in West Orange in 1982, Kelly (Figure 6.1) went on to study electrical engineering at the State University of New York Maritime College, graduating in 1987. From there, he transitioned directly into the U.S. Navy and qualified as a naval aviator two years later. After spending a short time at the Naval Air Station (NAS) in Beeville, Texas, Kelly made tracks for Fighter Squadron 101 at Oceana, Virginia, where he qualified in the F-14. With F-14 training under his belt, Kelly flew with Fighter Squadron 143 and made several deployments to the North Atlantic, the Red Sea and the Persian Gulf. For many, this would be enough adventure, but astronauts – or even fighter pilots for that matter – aren't wired like normal people. In 1993, Kelly was given the go-ahead to be trained as a test pilot at the Naval Test Pilot School for 18 months. While there, he put the F-14 and F/A-18 through their paces while developing new digital flight control systems and testing high angles of attack. During his time as a pilot, he logged more than 8,000 hours flight time across 40 aircraft, and made more than 250 carrier landings to boot. And amidst all the high-end flying, he found time to study for and complete a master's degree in Aviation Systems *and* submit his astronaut application. NASA snapped him up and in August 1996, Kelly reported for two years of astronaut training duty at Johnson Space Center (JSC). After completing training and a short stint in the Astronaut Office Spacecraft Systems and Operations Branch, Kelly was quickly assigned to flight status on the STS-103 mission as a pilot (Figure 6.2). STS-103 was the third Hubble Space Telescope servicing mission and was launched on December 20, 1999.

While on orbit, *Discovery* grappled Hubble at a record-setting altitude of 609 kilometers, and STS-103 crewmembers performed three spacewalks to replace gyroscopes, sensors

Figure 6.1.   Scott Kelly in the Cupola. Credit: NASA

and computers. Following his return from the very successful servicing mission, Kelly was assigned as NASA's Director of Operations at the Gagarin Cosmonaut Training Center in Russia, where he also trained as a back-up crewmember for Expedition 5. Following that assignment and a spell as Space Station Branch Chief in the Astronaut Office, Kelly was assigned as a crewmember on board NASA's underwater research facility 20 meters below the surface of the Atlantic Ocean, as part of the five-day NEEMO 4 (NASA Extreme Environment Mission Operations) expedition that took place in September 2002 (he also served as a crewmember on NEEMO 8 in April 2005). Following NEEMO 4, it was time for Kelly to begin training for his first command assignment – an assembly flight, no less – STS-118. On August 8, 2007, *Endeavour* was launched with Kelly in command, and two days later the Shuttle arrived at the ISS following a flawless docking with Pressurized Mating Adapter 2 (PMA-2).

During their visit to the ISS, Kelly's crew delivered the station's S5 truss segment, several tonnes of supplies and the External Stowage Platform 3. They conducted four spacewalks to install the truss and performed maintenance work on the P6 radiator. Following his return from STS-118, Kelly was nominated as a back-up crewmember to Expedition 23 and as a prime crewmember to Expedition 25/26. To that end, he started mission-specific training with his back-up colleagues, Russian cosmonauts Oleg Skripochka and Aleksandr Kaleri. Two years later, Kelly, Kaleri and Skripochka watched as the prime crew of Expedition 23 headed up to the ISS. With the prime crew on their

Figure 6.2.   Scott Kelly with the closeout crew, STS-103. Credit: NASA

way, Kelly and his crew switched to their own prime assignment and spent the next few months gearing up for their October launch to the orbiting outpost.

On October 7, 2010, Soyuz TMA-01M launched from Baikonur and, after the standard 34-orbit approach, docked with the ISS. Kelly, Kaleri and Skripochka joined the Expedition 25 crew, Shannon Walker, Doug Wheelock and Fyodor Yurchikhin. During their time on orbit, the Expedition 25/26 crew performed more than 100 experiments, docked multiple Progress vehicles, and conducted spacewalks, which pretty much describes routine day-to-day activities on board the station. One of many highlights of the mission was the arrival of the Shuttle *Discovery* on February 26, carrying a complement of six and delivering the now-famous Robonaut (Figure 6.3), together with the Multipurpose Module stuffed to the gills with equipment.

By the time Kelly set foot on Earth again on March 16, 2011, he had racked up 180 days in space across three missions. In November 2012, after a stint as the ISS Operations Branch Chief, Kelly was confirmed to the crew of a one-year mission to the ISS, together with Mikhail Kornienko. This meant he would serve across four missions: as Soyuz Flight Engineer on Soyuz TMA-16A and ISS Flight Engineer for Expeditions 43 and 44, and then as Commander of Expeditions 45 and 46. Finally, after nearly a year in space, Kelly would return to Earth on March 16, 2016 on Soyuz TMA-18M, which would be commanded by Sergei Volkov.

Figure 6.3. Scott Kelly with Robonaut, which was co-developed by NASA with General Motors, and was designed to serve as an assistant to ISS astronauts. Unlike the plain robots wheeling around on Mars, Robonaut is very humanoid in appearance, and there's a reason for that: NASA's little helper has been designed to manipulate any tool an astronaut can use, and that level of dexterity requires a design very similar to a human. The idea first came about in 1996, and the first iteration was born in 2000. Then, in 2006, NASA's Dexterous Robotics Laboratory at Johnson Space Center joined forces with GM to design Robonaut. Credit: NASA

Sergei Volkov

Sergei Aleksandrovich Volkov (Figure 6.4) is a second-generation cosmonaut. His father, Aleksandr Volkov, flew three missions between 1985 and 1992, racking up more than a year in space; a record the younger Volkov was intending to break during his time on the ISS. Like Kelly, Volkov comes from a military background, although his career was cut short when he was selected as a Test Cosmonaut Candidate in December 1997 at the age of just 24. Two years later, he passed his space training exams, was promoted to Test Cosmonaut and joined the Cosmonaut Corps, where he began training for a mission to the ISS. Less than two years after he began training, Volkov was confirmed to the back-up crew of Expedition 7, together with Sergei Krikalev and John Philips. When *Columbia* met with tragedy, however, their training was put on ice. Eventually, Volkov was re-assigned to the prime crew for Expedition 11 and resumed training for the STS-121 mission. But when he was replaced by European Space Agency (ESA) astronaut Thomas Reiter, Volkov was

Figure 6.4.   Sergei Volkov in Zvezda. Credit: NASA

once again re-assigned, this time to the group of cosmonauts slated for Expeditions 15, 16 and 17. Then, in February 2006, Volkov began training as back-up to Brazilian spaceflight participant Marcos Pontes, who flew on Soyuz TMA-8. Finally, three months later, Volkov transitioned to a firm flight assignment for Expedition 17 and began training with Oleg Kononenko and NASA astronaut Dan Tani. A year later, Volkov was confirmed as Commander of the mission that was due to launch early 2008.

On Volkov's Soyuz TMA-12 mission that launched on April 8, 2008, he and Kononenko were joined by Korean spaceflight participant Yi So Yeon (Figure 6.5). On their arrival at the station, they joined the resident crew of Peggy Whitson, Yuri Malenchenko and Garrett Reisman. Following Whitson's departure a week later, Volkov became the youngest person to command the station. During his tenure as commander, Volkov oversaw the comings and goings of the Automated Transfer Vehicle (ATV), the Progress M-64 vehicle, and the Shuttle *Discovery*, which brought the Japanese Kibo module to the station as part of STS-124. In addition to command duties, Volkov also got to venture outside the station in July to inspect the Soyuz TMA-12. The young commander conducted a second spacewalk later that same month to outfit the Zvezda module, in preparation for the arrival of the Russian Poisk module slated for the following year.

Figure 6.5. Yi So Yeon, Sergei Volkov, Oleg Kononenko, Soyuz TMA-12. During So Yeon's return on board Soyuz TMA-11, on April 19, 2008, together with Peggy Whitson and Yuri Malenchenko, a malfunction with the Soyuz resulted in the vehicle following a ballistic re-entry which subjected the crew to 10-G. As a result of the awry re-entry, the TMA-11 craft landed 420 km off-course. So Yeon was hospitalized on her return to Korea, due to severe back pains caused by spinal re-compression. Credit: NASA

On October 12, 2008, it was finally time for Volkov to depart the station after 198 days in orbit. After returning from his first trip to space, it wasn't long before Volkov was assigned to his next mission. This time it was Expedition 28/29, which meant he would train as back-up for Expedition 26/27. Once again, Volkov was confirmed as Commander. Within a year of his return from almost seven months on orbit, Volkov began mission-specific back-up crew training, with Mike Fossum of NASA and Satoshi Furukawa of the Japanese Space Agency (JAXA). In September 2010, the Expedition 26/27 prime crew launched and Volkov (and his crew) once again switched to prime crew training for what would be a June 2011 launch. Heading into space for the second time on June 7, 2011, aboard Soyuz TMA-02M, Volkov (together with Fossum and Furukawa) arrived at the station two days later, where he and his colleagues were met by the incumbent crew of Aleksandr Samokutyaev, Andrei Borisenko, and Ron Garan. During their increment on board the station, the crew witnessed the arrival of the Progress M-11M vehicle and the final flight of the Shuttle program, STS-135, when *Atlantis* delivered cargo.

Figure 6.6.   Mikhail Kornienko. UNESCO. December 2014. Credit: Russian Space Agency

On September 16, Borisenko, Samokutyaev and Garan departed the station, leaving the orbiting outpost temporarily short-staffed due to a faulty Soyuz rocket that delayed the next incoming crew. After a two-month delay, Soyuz TMA-22 launched on November 14, 2011, carrying Anton Shkaplerov, Anatoli Ivanishin and Dan Burbank. After a week of hand-over tasks, Soyuz TMA-02M departed the station, returning Volkov to Earth after a mission that had lasted 167 days. When added to the length of his first mission, Volkov had now racked up 365 days in space. A full year, but still short of the 391 days logged by his father. Close but no cigar. Volkov wouldn't have long to wait for family bragging rights over that duration record though, because in December 2012 he was assigned as back-up to Mikhail Kornienko, who was in mission training with Scott Kelly for the one-year increment. Once Volkov was no longer required as back-up, he transitioned to the role of Flight Engineer as part of the Expedition 45/46 crews.

Mikhail Kornienko

In common with many of the Expedition 45/46 crew, Mikhail Kornienko (Figure 6.6) transitioned to the business of manned spaceflight via a military career, spending several years in the Russian Army before working as a militiaman for six years. In 1987, he graduated from the Moscow Aviation Institute with a degree in Mechanical Engineering, with a specialization in liquid propellant rocket engines. For the next several years, he worked at the General Machine Building Design Bureau in the technical department, before transitioning to work at RSC Energia in 1995 where he worked on cosmonaut training for spacewalks. Three years later, he was selected as a Cosmonaut Candidate and was cleared for basic training, which he completed in December 1999. Between August 2001 and the beginning of 2003, Kornienko focused on mission training for long duration increments on board the station, and also logged training increments for a possible Shuttle flight. Originally assigned to the Expedition 8 back-up crew, Kornienko was bumped from that flight and had to wait until 2006 before he received his next assignment, when he was confirmed as back-up to the Expedition 15 crew. Two years later, he was assigned to the prime crew for Expedition 23/24, together with Commander Aleksandr Skvortsov and Tracy Caldwell-Dyson. Fortuitously, as it would turn out, Scott Kelly had been assigned as a back-up crewmember to Expedition 23/24, which meant the future year-long mission candidates got the chance to spend time training together.

Finally, after the best part of 12 years waiting for his flight – almost as long as Canadian Space Agency astronauts have to wait! – Kornienko lifted off in Soyuz TMA-18 on April 2, 2010. Once on the station, Kornienko, Skvortsov and Caldwell-Dyson joined Expedition 22 crewmembers Oleg Kotov, Soichi Noguchi and Timothy Creamer. Three days after Kornienko's arrival, the Shuttle *Discovery* arrived at the ISS on STS-131, delivering science racks and crew quarters to the station. It was a busy mission that featured three spacewalks to install the S0 Truss and replace a balky gyroscope. After *Discovery's* departure on April 17, the crew only had to wait another two weeks before their next visit, by the Progress M-05M vehicle that delivered 2,588 kilograms of cargo.

As with every station increment, Kornienko's mission was a busy one, featuring the comings and goings of Progress vehicles, Shuttle visits, the commissioning of new crew accommodation in the Harmony module, and the installation of new modules such as the Rassvet docking port and research module, which was hauled up by *Atlantis* during STS-132 in May. The next to visit after *Atlantis* was Soyuz TMA-19, which launched on June 15, carrying Fyodor Yurchikhin, Shannon Walker and Doug Wheelock. This was good news for Kornienko's crew because it meant the station complement was back up to six. In addition to all the assigned tasks, which included routine rendezvous and docking operations, science, and myriad maintenance tasks, the crew also had to deal with what must have seemed a never-ending list of malfunctions. These don't hit the headlines much anymore because of the public perception of manned spaceflight being routine. As any astronaut who has spent time on board the station will tell you though, it is anything but. Systems break down all the time, and when that system happens to be a life support system, or a sub-system, or a sub sub-system, then that breakdown raises the very real specter of a loss of mission (LOM) scenario. Which is what station crews face on a daily basis.

In July, the crew found themselves battling with the failure of the oxygen generation system and a balky water recovery unit. While they were trying to fix those, they also had to slot in a spacewalk to install automated rendezvous antennas for future visiting vehicles. After the oxygen generating system and the water recovery unit problems, the crew was faced with yet another life support system failure when the ammonia pump module failed. This system circulates ammonia between heat exchangers as part of the station's cooling system: a failure of this system means the station gets warm *very* quickly. To limit the heat buildup, the crew powered down several ISS systems and began preparing for a series of emergency spacewalks.

Figure 6.7. Tim Kopra in the Neutral Buoyancy Laboratory. Credit: NASA

It took three spacewalks – performed by Wheelock and Caldwell-Dyson – to remove ammonia and electrical lines from the balky pump, replace the pump from the station's supply of spares (note to those planning a manned Mars mission: you will need lots *and lots* of spares) and then go about the lengthy process of reconnecting all the cables and lines. Finally, on August 21, three weeks after the issue surfaced, the ammonia pump problem was resolved. As August drew to a close, Kornienko's crew began preparations for their return to Earth. On September 22, Aleksandr Skvortsov handed over command to Wheelock, and two days later the Expedition 24 crew bundled into the Soyuz TMA-18 and headed for home. Kornienko had logged 176 days on orbit. On his return, Kornienko didn't have to wait long before being assigned to his next mission. In November 2012, it was announced that he was to fly the year-long mission to the station together with Kelly. Which is where we pick up the Tim Peake story again.

Tim Kopra

Joining Peake and Yuri Malenchenko in the cramped confines of the Soyuz (Figure 6.7) was Tim Kopra, a colonel in the U.S. Army who had previously flown as Flight Engineer on board the ISS in 2009. After graduating from the U.S. Military Academy, Kopra began a three-year assignment at Fort Campbell, Kentucky, before transitioning to the 3rd Armored Division in Germany in 1990. While stationed in Germany, he was deployed to the Middle East in support of Operations *Desert Shield* and *Desert Storm*, before returning to the U.S. in 1995 where he completed a master's degree in Aerospace Engineering. He then followed a path that echoed Peake's, starting work as a test pilot and helicopter pilot instructor. That assignment lasted until 1998, when Kopra started work at NASA as a vehicle test engineer on the Shuttle and ISS. Two years later, he was selected as an astronaut in NASA's Group 18. After completing the standard two years training, Kopra began work in the Space Station Branch, where he worked testing crew interfaces on the U.S. ISS modules.

While waiting for his first flight assignment Kopra, together with Nicole Stott and Maksim Suraev, completed survival training in Russia – in January; great preparation in case the Soyuz landed in Siberia. Like Kelly, Kopra also served as a crewmember on board NEEMO when he worked in Aquarius in September 2006. He was then assigned as back-up to Garrett Reisman and Gregory Chamitoff for Expedition 16/17, after which he assumed the position of prime exchange crewmember, to launch aboard STS-127. Finally, after nearly ten years as an astronaut waiting to fly, Kopra had his chance when he blasted off on board *Endeavour* on July 15, 2009 to replace Koichi Wakata on the ISS resident crew. During his time on the station, Kopra had the opportunity to perform a spacewalk, assist with the deployment of the Exposed Facility of the Japanese Experiment Module, support myriad science activities and conduct the routine maintenance tasks that form part of day-to-day life on board the orbiting outpost. When *Discovery* docked with the station during STS-128 on August 31, 2009, Nicole Stott took Kopra's place, and Kopra returned to Earth after 58 days in space.

Kopra barely had enough time to adjust to regular Earth gravity before he was assigned to STS-133, but fate intervened. Just weeks before the launch, a bike accident sent Kopra to the hospital with a hip fracture. His place on the crew was taken by Stephen Bowen, who flew to the ISS in March 2011. Following his recovery, Kopra was confirmed to the Expedition 46/47 crew and began mission training with fellow crewmembers Peake and Malenchenko.

Yuri Malenchenko

Born in 1961, Yuri Malenchenko (Figure 6.8) is a colonel in the Russian Air Force. After graduating from the Kharkov Military Aviation School in 1983, he worked as a pilot, senior pilot and flight lead at the Zhukovsky Air Force Engineering Academy until 1987, before being selected as a cosmonaut. After completing cosmonaut training in 1989, Malenchenko began more advanced training for spaceflight. He trained as a back-up for the *Mir* 14 and *Mir* 15 missions, before being selected as a prime crewmember for the *Mir* 16 mission that flew in 1994. After completing two spacewalks and logging 126 days in space, Malenchenko returned to terrestrial work as a cosmonaut instructor, before being assigned to Shuttle flight STS-106. To prepare for his flight, Malenchenko trained at the JSC from October 1998 to September 2000, before launching on board *Atlantis* for a 12-day space station assembly flight (ISS-2A.2b). After his Shuttle flight, the next logical step was a long duration increment on board the ISS, which was duly confirmed shortly

Figure 6.8. Yuri Malenchenko. Credit: Russian Space Agency

after Malenchenko's return. He began training for his ISS increment in January 2001 and launched on April 26, 2003. Malenchenko racked up another 185 days in space before returning on October 27, whereupon he resumed his work as a cosmonaut trainer. Then it was back to flight training for yet another flight, as a back-up to Expedition 14 before assignment to the prime crew of Expedition 16. This mission, which was dedicated to assembly, maintenance and science operations, launched on board Soyuz TMA-11. Following an eventful ballistic re-entry, Malenchenko's return after another 192 days on orbit gave him an accumulated total of 514 days in space, a period that had included four spacewalks.

DOWN TO WORK

The Expedition 46/47 crew didn't have to wait long before having to deal with the first of what would be many arrivals and departures at the orbiting outpost. First on the timeline was the undocking of the Progress M-28M spacecraft on December 19. The Progress freighter (see sidebar and Figure 6.9) had arrived at the station on July 25, 2015, when it

Figure 6.9. Progress cargo vehicle. Credit: NASA

had delivered three tonnes of food, fuel and supplies to the crew. Now it was time for the vehicle to undock to make space for another Progress. The reason for this is that the ISS only has a limited number of berthing ports.

Progress MS Cargo Ship in a Nutshell
The Progress MS freighter is the latest in a long series of Progress spacecraft, the roots of which can be traced back to 1978. The unmanned cargo ship is based on the manned Soyuz vehicle, and is capable of carrying cargo – in its pressurized carrier – as well as propellants and water. It is launched on the Soyuz rocket and can dock at any port on the Russian Segment of the ISS. Once the vehicle is secured on the station, the crew opens the hatch to the pressurized compartment and begins unloading the supplies. After dealing with the cargo, the crew turns their attention to the gases that are released to help repressurize the station, and the propellant that is transferred using a special system that routes the fuel into tanks on the Russian Segment. Once the vehicle is empty, the crew loads it up with trash before closing the hatch and undocking the spacecraft, in preparation for its fiery and destructive re-entry into Earth's atmosphere.

Table 6.1a Progress M Technical Details

Type	Progress M
Length	7.23m
Maximum Diameter	2.72m
Cargo Module Diameter	2.20m
Span	10.6m
Launch Mass	7,200kg
Cargo Volume	6.6m³

Table 6.1b Instrumentation and Propulsion Module

Diameter	2.72m	Oxidizer	Nitrogen Tetroxide
Launch Mass	~2,900kg	Fuel	Unsymmetrical Dimethylhydrazine
Habitable Volume	0m³	Propellant Mass	880kg
Main Propulsion System	KTDU-80	Power Generation	2 Solar Arrays
Main Engine	S5.80	Span	10.6m
Trust	2,950N	Thrust	26.5N/130N
Attitude Control	28 DPO Thrusters	Power	1,000W

Table 6.1c Cargo Module

Total Payload	2,350kg
Maximum Dry Cargo	1,800kg
Water	420kg
Air/Oxygen	50kg
Refueling Propellant	880kg
Disposal Cargo	Up to 1,600 kg
Liquid Waste Capability	400kg

Progress

In common with the Soyuz, the Progress (see Appendix V and Table 6.1a) comprises three sections; the Instrumentation and Propulsion Module (Table 6.1b), the Refueling Module (this takes the place of the Entry Module in the manned Soyuz configuration) and the Pressurized Cargo Module (Table 6.1c), which features the docking system and the propellant transfer system. At launch, the vehicle weighs up to 7,200 kilograms and it can be loaded with up to 1,800 kilograms of dry cargo, 420 kilograms of water, 50 kilograms of air and 850 kilograms of propellants.

On its approach to the station, the Progress makes use of its 28 multidirectional attitude control thrusters, controlled by the TsVM-101 flight computer and MBITS telemetry system housed in the Instrumentation Module. During the course of its rendezvous with the station, the Progress executes a series of adjustments, called *phasing burns*, to refine its attitude and approach. This happens while the vehicle is still some distance from the station. As the Progress closes in on the station, the on board KURS rendezvous system communicates with the KURS-A on board the ISS to exchange navigation data. As this information is updated, the Progress performs braking and course maneuvers until it finally arrives at a point 400 meters from the station. At this stage, the crew takes over and remotely pilots the Progress using the TORU system. As the Progress inches closer, it begins a fly around to prepare for alignment with the docking port. At 200 meters from the ISS, the vehicle begins a period of station-keeping that allows the Mission Control teams and the astronauts on board to check alignment and the health of on board systems. Once all systems are in the green, the Progress pulses its thrusters to power the vehicle along at just 10 centimeters per second until it executes a soft docking. Following the soft docking, hooks are used to form a hard mate. Then a one-hour leak check is performed, after which the crew can finally get their hands on the cargo. Any propellant not used for rendezvous and docking operations (usually between 185 to 250 kilograms) is stored for station reboost operations, which are required every few months since the station loses about two kilometers of altitude every month. In reboost mode, the Progress points four or eight of its attitude control thrusters in the desired direction to increase the station's orbit (the thrusters have to be used because the main engine puts too much stress on the station).

Unplanned EVA

With one Progress dispatched and on its way to a fiery re-entry in Earth's atmosphere, it was time to prepare the station for the next incoming cargo vessel. Progress 62 (Progress MS 431), which launched on top of a Soyuz 2-1A on December 21, was carrying 2.8 tonnes of food, fuel and supplies. While the Progress was on its way, Kelly and Kopra conducted an unplanned spacewalk to move the station's transporter rail car, so it could be fitted in place in preparation for the incoming Progress. The transporter had stalled five days earlier, just centimeters from the point it started its trip across the worksite. Kelly and Kopra's job was to check the position of the brake handles to make sure the transporter could be moved back to the worksite by the controllers at Mission Control. Fortunately, the fix was a simple one and the two spacewalking astronauts were able to release the brake handles. They spent their remaining time outside completing other housekeeping tasks, including routing cables in advance of the installation of a docking adapter. The short, three-hour 16-minute spacewalk was the 191st in support of the ISS, for those who like to keep track of these stats. For Kelly and Kopra, it was just another routine day on orbit. For Peake, it was an opportunity to follow the routine of the two spacewalkers, mindful of the fact that he was due to make his own trip outside the station the following month.

With Kelly and Kopra's spacewalk done and dusted, the next major event on the agenda was the incoming Progress, which was due to arrive on December 23. After completing its 34 orbits of the Earth (normally this would have been a six-hour trip, but Russian managers wanted to allow extra time to test new systems fitted on the vehicle), the freighter made its approach to the ISS. This Progress – the first of the MS category – was a slightly different version to the one that had just departed thanks to a number of upgrades, including an external compartment to carry small satellites, enhanced micrometeoroid and orbital debris (MMOD) protection, updated communication links, and a new camera to help crewmembers manually dock the vehicle if the automated system failed. The Progress arrived at the ISS two days later, and was docked with the station's Pirs docking compartment. The Progress arrived at the station's Pirs module at 10:27 GMT, and after docking, the crew went to work unloading the supplies (Table 6.2).

Table 6.2 Progress 62 (MS 431) Manifest

Propellant	880 kilograms
Water	420 kilograms
Oxygen	24 kilograms
Air	22 kilograms
Russian food rations	379 kilograms
Napkins	199 kilograms
Russian power supply system parts	162 kilograms
Medical and hygiene gear	60 kilograms
Batteries, hard drives	35 kilograms

Christmas on orbit

"Christmas is traditionally a time for friends and families to get together. And although we can't be with our friends and families this year, we'll be orbiting the Earth 16 times on Christmas Day, and sending all our good wishes to everybody back down on beautiful planet Earth."

Tim Peake, from ISS

The next major event on orbit was Christmas. Christmas in space is a tradition that stretches back all the way to Apollo 8, when Jim Lovell read from the Book of Genesis as the vehicle circled the Moon. The tradition was continued in the Skylab years, when the crews built a Christmas tree from used food cans, and the winter holiday season became a steady tradition on board the ISS as soon as crews began permanent occupation of the outpost. It was no different in 2015, with the crew decorating the station with a small tree and stockings. A few days before Christmas, Peake had attempted to call his parents but had to leave a voicemail. When he tried again the day before Christmas he didn't have any better luck, because he dialed the wrong number (see sidebar). Trying to make amends, Peake tweeted an apology that made the rounds in the popular British press:

Tim Peake
@astro_timpeake
I'd like to apologise to the lady I just called by mistake saying 'Hello, is this planet Earth?' – not a prank call ... just a wrong number!
6:04 PM - 24 Dec 2015

"Hello, is this planet Earth?"
That is what retired schoolteacher, Betty Barker, heard on Christmas Eve, when a certain British astronaut tried phoning home. Somehow, Major Peake has dialed the wrong number, proving that even highly-trained astronauts sometimes make mistakes. Barker, who lives in the same area code as Peake's family, thought Peake was a crank caller or someone who had been on a drinking binge, so she said 'No'. The next day, her husband, Patrick, heard the news, and before she knew it, Betty was being interviewed by the *Daily Mirror*.

Figure 6.10.   The Neutral Buoyancy Laboratory. Credit: NASA

WALKING IN SPACE

The low Earth orbit miscommunication was quickly forgotten as Peake and Kopra set their sights on the first spacewalk of 2016. Spacewalk training is something best practiced in a tank of water. A huge tank of water, as it happens. And the mother of all tanks happens to be in Houston. It's called the Neutral Buoyancy Laboratory (NBL, Figure 6.10), and is where astronauts spend hundreds of hours practicing for spacewalks by being neutrally buoyant while swimming around a mock-up of the ISS. Being neutrally buoyant is not quite the same as being weightless in space, but it's a useful approximation.

Practice makes perfect

For each spacewalk, astronauts must practice, practice *and practice*, not just to ensure that the actual EVA is executed perfectly, but also to train for contingencies. But before the EVA training cycle begins, an astronaut has to demonstrate some aptitude for the job. Many people assume that anyone lucky enough to be selected to work on orbit is automatically qualified to do the job of a spacewalker, but that isn't the case. In the ESA training flow, all candidates start by attending practical classes that cover the use of EVA tools, before being introduced to the nuts-and-bolts of what constitutes working outside a vehicle in space. These classes introduce the candidates to the use of tethers and restraints, and the skills needed to communicate while performing maintenance tasks 400 kilometers above the surface of the Earth. Once the classroom is checked off, astronaut candidates move on to the water phase, which is held in the NBL. Here, each water training session gradually builds on skill-sets developed in the previous session, with the aim of teaching the students to be comfortable in a neutrally-buoyant environment. To achieve this goal, students are taught skills to improve their situational awareness, how to optimize body position and how to conduct correct lifeline protocols. It is during these sessions that candidates realize just how complex the task of spacewalking is – a heady combination of physical exertion, mental fortitude, supremely well-balanced psychomotor skills, and finely tuned hand-eye coordination.

Not surprisingly, some candidates fall by the wayside, but those who demonstrate an aptitude for this very unique task are invited to sign up for the EVA Task Skills Program. This program comprises a series of ever-tougher training sessions designed to improve an astronaut's ability to support EVA operations. A typical phase begins with a table-top discussion, followed by task-specific scuba sessions and finally the NBL suited events. During their initial scuba sessions, astronauts must learn how to navigate along the trusses to various worksites. Once they are proficient in getting around the station, they are taught how to evaluate what tools and tethers might be required for each task and also to estimate how long various tasks might take. Gradually, as the astronauts develop the wide range of skill-sets required for working in the vacuum of space, they acquire the necessary situational awareness. As the training evolves, astronauts spend more time on mock-ups submerged in the NBL, with coaches providing critique and feedback. And once the training is complete, astronauts are deemed qualified for an EVA assignment on a mission, which means spending a lot more time submerged in the NBL

> "For each specific spacewalk, there are several training units to be completed. One EVA run lasts around five hours, and the standard right now is that you spend five to seven times as long in the NBL at Houston for each EVA, depending on the difficulty. In addition to that, you train a lot of contingency scenarios."
>
> *Christer Fuglesang, ESA Astronaut.*

NASA's NBL is unlike any other pool on the planet. A friend of mine used to work as an EVA flight controller at NASA and I had the opportunity to visit. Just standing at the edge of the pool provokes a sense of vertigo because the body of water is just so damn big. A staggering 23.5 million liters of water is confined by a tank measuring 62 meters in length, 31 meters in width and 12.34 meters deep. This is the largest indoor body of water in the world. Under the surface, it's easy to make out full-scale mock-ups of various ISS modules, as well as those of visiting vehicles such as SpaceX's Dragon. While it is the biggest pool in the world, however, even the mighty NBL can't house a full-scale mock-up of the ISS. To give you an idea of the size of the station, consider the following: the ISS comprises 11 trusses and the NBL can only accommodate three of them. As with any pool, the smell of chlorine permeates the air and there is the white noise of pumps doing their thing. PA announcements keep everyone on track with what's going on, as support divers (the NBL has 28 of them) scurry around attending to the day's tasks.

Figure 6.11.   Tim Peake in his spacesuit. Credit: ESA

Some might wonder why so many spacewalks must be conducted now that the ISS is complete, but the reason is simple: repair and maintenance. Solar arrays get jammed, power units fail and bits and pieces need to be replaced. Some of these tasks can be performed by the robotic arms attached to the ISS, but others need the attention of space-walking astronauts. And those astronauts need to train, train and *train* to ensure every movement is executed perfectly, because their working environment is lethally hostile and because doing anything on orbit is expensive. For example, the cost per crewmember per day runs to $7.5 million, so one spacewalk lasting six hours costs close to $2 million.

But let's get back to the NBL. It's a few minutes after eight in the morning and the stars of the show turn up for the day ahead. They have likely been awake for hours. First on the schedule is the mission briefing, which was a lot shorter than I expected although perhaps not surprising given that everyone here is a supremely well-trained and well-practiced professional, each of whom has done this EVA business many times before. Just another day at the office for these guys, I guess. After the Test Coordinator has outlined the dive ahead (using a PowerPoint incidentally), the dive team lead takes over and reminds the safety and camera divers about various safety issues.

Dressed for the occasion

At the core of EVA training is the very impressive and extraordinarily expensive ($12 million) EMU suit (see Figure 6.11 and sidebar). Given the cost of this suit, you might think it's a breeze to don, but this is science fact and not science fiction. Forget all those scenes in the movies that show astronauts slipping into their spacesuits in the time it takes to say, 'liquid cooling garment'. The early 21st Century reality is of an astronaut and a suit engineer spending the best part of a quarter of an hour pulling up the pants, shrugging into the upper torso unit and connecting the ventilation garment. Once that is done, there are myriad hookups that must be connected, the umbilical cable to be taken care of and then finally the donning and sealing of the helmet. Once the process of suiting up is complete, the astronauts turn their attention to the tools and equipment they will be using during the sim, while the support divers busy themselves by applying toothpaste – yes, toothpaste! – to their masks to prevent fogging.

Custom-made spacesuits

The global market for pressure and spacesuits is a small one. In 2015, just 50 suits were sold globally for a total market value of $100 million. The point is that the business of building suits is a small and very unique one. You can't go on E-Bay to buy these things. Each astronaut is measured for their NBL suit and a flight suit, a task that involves 37 anthropometric measurements of the body and another 40 of the hands. These stats are then run through a computer program that spits out a sizing sheet, which is then used to pick off-the-shelf suit elements. One component that doesn't require sizing is the space diaper, although this being NASA, it has an acronym: The Maximum Absorbency Garment, or MAG. The MAG is the first item of clothing the astronauts don, followed by the Thermal Cooling Unit, or TCU, their thermal underwear, and then the Liquid Cooling Ventilation Garment (LCVG). Then it's time to slip into the suit. All 145 kilograms of it! One of the bulkier components of the suit is the chest-mounted Display and Control Module (DCM). The DCM is about the size of a bread box and sports oversized controls that can be manipulated by astronauts wearing chunky gloves. The DCM, which functions in much the same way as the dashboard in your car, has gauges and controls, with all the labels written backwards. This is because the labels are not in line-of-sight, but instead must be read by the astronaut using a wrist-mounted mirror. A significant contribution to the weight of what is in effect a small spacecraft is the Primary Life Support System, or PLSS. Think of this as the engine of the spacesuit. Another major feature is the metal harness that is termed the mini-workstation, which allows the astronaut to carry an assortment of tools, and there is also the Body Restraint Tether (BRT), which is the large cable that the spacewalker uses to attach themselves to the ISS structure.

The support team

Key to the precisely choreographed exercise that is a simulated spacewalk are the safety divers (Figure 6.12). These guys have all sorts of diving backgrounds. Some are ex-SEALs, some cave divers, and some retired engineers whose passion just happens to be diving. Regardless of their disparate backgrounds, each of them has a logbook filled with hundreds if not thousands of dives. This is *not* the sort of job you can apply for with a simple

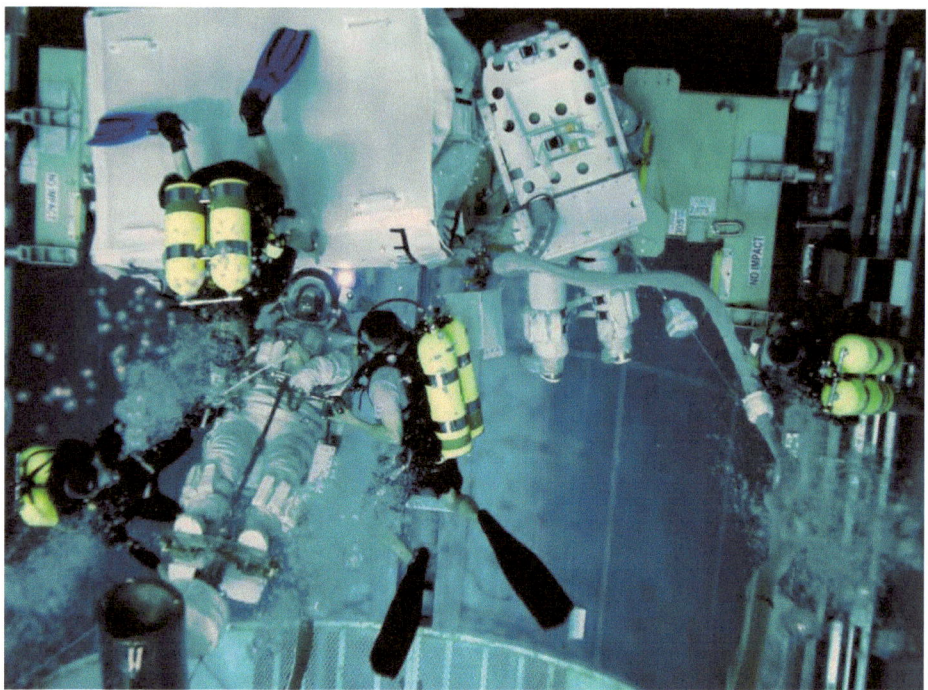

Figure 6.12. The Neutral Buoyancy Laboratory support team divers. Credit: NASA

PADI certification. A safety diver must have enough experience to dive on autopilot, an instinct that allows them to concentrate on the job at hand and not be occupied with their own diving. Breathing nitrox[1], a 46 to 54 percent mix of oxygen and nitrogen, safety divers wear standard scuba gear, although the stuff takes more of a beating than the equipment used by recreational divers: when they aren't training astronauts, the safety divers are invariably cleaning and maintaining equipment.

Another key person in the NBL is the EVA Task Group lead, whose job it is to plan submerged spacewalks. Each individual simulated spacewalk is the end product of dozens of planning meetings, in which the core elements of each sim are broken down into their component parts, analyzed, broken down some more, and then slotted into the timeline. The EVA Task Group lead, together with a small army of Test Coordinators (TCs), is responsible for orchestrating the sim procedures so that the whole operation progresses seamlessly. During the sim, the TCs serve a dual role as flight controllers, in much the same way as the flight controllers in Mission Control. While the sim is underway, the TCs watch

---

[1] The astronauts breathe the same nitrox mix as the divers, but the pressure inside the EVA suits (4.3 psi) must be carefully regulated in response to the changing pressure as astronauts move up and down the water column. This is more difficult than it sounds, because the water exerts 0.5 psi per foot of depth so there might be a pressure differential of 2 psi if an astronaut is vertical. To compensate, the flow rate of gases to the suit is varied as the pressure changes.

the astronauts like hawks, making sure their movements are consistent with what will be required on orbit. For example, the simple task of traversing a truss structure ('translation', in NASA parlance) requires a whole new understanding of how to apply simple physics, and the TCs are there to offer advice on how to do exactly that. In addition to training astronauts for upcoming missions, the TCs may occasionally work with astronauts to troubleshoot a spacewalk that went awry. In such a situation, astronauts and TCs watch video of the spacewalk, troubleshoot it and then the astronauts jump into the NBL, as stand-ins for their colleagues on orbit, to try and figure out a better way to conduct the EVA.

Rounding out the NBL team are the Test Directors (TD), who monitor the TCs and also the myriad systems required to ensure a sim in the NBL progresses without a hitch. There is a TD that monitors the engineering systems, another whose job it is to make sure the life support systems stay in the green, and another who makes sure that the 22 pan-and-tilt cameras and floating cameras are providing the feedback required to run the sim. Given the amount of monitoring that must be done, it's not surprising that TDs are surrounded by banks of video monitors that show ... well, pretty much everything that is going on in the sim. If there is a problem during the execution of the sim, then it's the job of one of the TDs to resolve it. This could be anything from a leak in a suit to a case of decompression sickness (DCS), which might result in symptoms such as nausea and/or loss of coordination.

DCS is the result of inadequate decompression following exposure to increased pressure. During a dive, body tissues absorb nitrogen in proportion to the surrounding pressure. As long as the safety divers remain at pressure, the dissolved nitrogen in tissues and blood presents no problem. But if the pressure is reduced too quickly, the nitrogen comes out of solution and forms bubbles in the tissues and bloodstream. For the NBL safety divers, DCS is a condition that could occur at the end of a dive when ascending to the surface. To avoid DCS, the safety divers ascend slowly (a hyperbaric chamber is available at the NBL for the immediate treatment of a diving-related DCS, although such an event has never occurred).

Astronauts must also take precautions to avoid DCS that could occur when going on spacewalks. The pressure inside the spacesuits that the astronauts wear on spacewalks is significantly lower than the ambient pressure of the ISS. For this reason, astronauts go through a denitrogenation process called a pre-breathe (Table 6.3) prior to every spacewalk.

Today's ISS astronauts perform what is termed the ISLE (In Suit Light Exercise) EVA pre-breathe protocol, developed during the final days of the Shuttle Program. The ISLE is an improvement on the older pre-breathe protocol, because it takes less time and also saves on valuable life support consumables. This is how it works. On the day of the spacewalk, the EVA astronauts don oxygen masks and the airlock pressure is reduced to 10.2 psi. The astronauts then don their spacesuits and the airlock pressure is brought back to 14.7 psi. The astronauts then spend the next 50 minutes pre-breathing in their suits before beginning a 50-minute light exercise program (flexing their knees for 4 minutes, followed by 1 minute of rest) while wearing their suits. At the end of this series of steps, the nitrogen in the blood is flushed out and the astronauts are ready for their tasks outside the station.

Table 6.3 Pre-breathe Protocol for ISS

Pre-breathe Protocol Steps	Ambient Pressure (kPa)	Start Time (min)	End Time (min)	Breathing Gas (% of O2)
1. Oxygen mask is donned to start denitrogenation	101.3	0	30	100
2. Decompression from 101.3 to 70.3 kPa	70.3	30	60	100
3. 8 hrs. 40 min. living at 70.3 kPa	70.3	60	580	26.5
4. Recompression from 70.3 to 101.2 kPa	101.3	580	590	100
5. Stay at 101.3 kPa for hygiene break	101.3	590	620	100
6. Decompression from 101.3 to 70.3 kPa	70.3	620	650	100
7. Don spacesuit	70.3	650	710	26.5
8. 50 min. of in-suit pre-breathe with 100% oxygen	70.3	710	760	100

A reason to go out

Peake's spacewalk had not been a part of the Expedition 45/46 mission plan until November 13, 2015. That was when a Remote Bus Isolator inside a Direct Current Switching Unit (DCSU) tripped, which resulted in a short in the Sequential Shunt Unit (SSU). The SSUs are an integral part of the station's Solar Arrays. The arrays feed power to these units, which in turn manage the voltage output of the array by using a series of capacitors that regulate power downstream of the SSUs. When telemetry showed a short inside SSU 1B on November 13, due to a burst of current, the DCSU responded as it was designed to do and tripped, to prevent equipment downstream from being damaged by the excess current. The outcome of the DCSU tripping was the loss of one of the eight power channels. While the station can function with just seven power channels, the condition is less than optimal because redundancy is lost in other systems. Hence the need for a spacewalk to resolve the problem. The problem with servicing an SSU is that they can only be replaced when the solar array is not generating power, which means the work can only be performed during an orbital 'night'. To maximize the time the crews can work when there is no power being generated from the solar arrays, mission managers target a time when there is a low beta angle, which is when the solar arrays are side-on to the Sun. The next window with a low beta angle was between January 12 and 18; hence the timing of Peake and Kopra's spacewalk on January 15.

Preparation for the spacewalk began on January 4, when Kopra checked out Extravehicular Mobility Unit (EMU) 3008 to make sure all the systems were functioning as expected. One of those checks involved working with the ground team to ensure that the EMU's UHF communication system and audio system worked. Later, Kopra configured suits 3005 and 3008 for 'loop scrubbing'. This sounds like a simple system check, but it is one of the most time-consuming maintenance tasks performed on the EMUs while in orbit. First, an ionic and particulate filtration (the actual 'scrubbing') is performed, followed by a 2-hour biocidal iodination procedure. This latter procedure is also conducted on the airlock cooling water loops. The next step is to dump and then fill the EMU feedwater tanks (a feedwater sample is stored for ground analysis). Once those checks are completed, the crew generally move on to more standard checks, such as EVA Cuff Checklist Procedures, resizing activity (to make sure the arm rings are the correct fit for each astronaut), removing gas from the payload water reservoir, filling the Liquid Cooling and Ventilation Garment (LCVG) and performing tether inspections. Preparation for a spacewalk is a busy time and it is just a snapshot of life on board the ISS. And on the subject of life on board the station, it is perhaps helpful to follow the activities of the crew through just one day. We'll take January 4, 2016. For those who are interested in the day-to-day activities on the ISS, NASA posts a useful blog at the following address:

https://blogs.nasa.gov/stationreport/

A DAY IN THE LIFE

We'll begin with Scott Kelly, who began his workday following his wake-up call by collecting saliva, blood and urine and placing the samples in the Minus Eighty Degree Celsius Laboratory Freezer for ISS (MELFI). With that job out of the way, Kelly swallowed a sodium bromide tracer as part of another experiment. Throughout the day, Kelly collected more urine, blood and saliva samples, but his main focus on January 4, 2016 was to support the fluid shifts experiment. This study investigates the effect of fluid shift on intracranial pressure and how this contributes to visual impairment in astronauts. It's a serious problem, because more than half of NASA's astronauts have developed visual impairments following long duration spaceflights. These visual impairments have included structural abnormalities in the eye (papilledema, choroidal folds, optic nerve sheath dilation, optic nerve kinking, and globe flattening) and elevated intracranial pressure. One of the theories that has been proposed to explain these changes is the fluid shifts toward the head that occur as a result of the absence of gravity. But like any theory, the hypothesis had to be tested, which is why ISS crewmembers are involved in various tests to assess metrics such as flow parameters, ocular pressure and changes in intracranial pressure. Astronauts are also used as guinea-pigs to test countermeasures – such as Lower Body Negative Pressure (LBNP, Figure 6.13) – to the Visual Impairment, Intracranial Pressure syndrome (VIIP, see sidebar). One of the many tests that comprise the fluid shifts investigation is the dilution measures collection (a way of measuring fluid compartmentalization indirectly by analyzing saliva and urine), which was one of Kelly's tasks that day.

Figure 6.13. Scott Kelly sports the Russian Chibis suit. The Chibis, a Lower Body Negative Pressure (LBNP) device, generates measured orthostatic stress to the cardiovascular system. Getting into the contraption is simple. First, the crewmember places both legs inside, making sure the pressure seal is at the iliac crest. Next, a microprocessor sucks out some air, creating a partial vacuum on the legs. As the suit does its thing, the crewmember remains connected to the station by an air hose and an electric cable. This cable is fairly long, enabling the crewmember to retain almost complete freedom of movement. Credit: NASA

VIIP and why it's so important
When the VIIP problem surfaced in 2005, it was the catalyst for researchers performing a retrospective survey of questionnaires that were submitted to 300 Shuttle astronauts. The analysis of the responses revealed that 23 percent of Shuttle astronauts had experienced near-vision changes and 11 percent had experienced post-flight changes. Following these findings, a study of long duration astronauts was performed to get an idea of the visual symptoms and physiological changes that were occurring during a typical four- to six-month ISS mission. The results were striking. Five crewmembers had disc edema and globe flattening, three had cotton wool spots, five had problems with near vision, and five had choroidal folds. Further investigation revealed a constellation of symptoms, including kinked optic nerves, scotoma, refractive deficits, optic disc protrusion, and elevated intracranial pressure (ICP)[*]. Some crewmembers, all of whom were males between 45 and 55,

experienced symptoms so severe that they had difficulty reading checklists. Something had to be done. After all, there was no way NASA could plan asteroid missions and expeditions to the Red Planet when crews ran the risk of becoming half blind! The first step was to establish a theory that explained the VIIP syndrome.

* If you're interested in this subject, the go-to reference guide is a Springer Brief titled: *Microgravity and Vision Impairment in Astronauts* by the author, published in 2015.

Several theories abound for VIIP. Some researchers reckon an increase in ICP is to blame. It sounds like a reasonable theory. After all, one of the first changes that occurs on entering orbit is a fluid shift to the head in the order of two liters. All that fluid rushing to the head is bound to cause an increase in pressure. The problem is that ICP sufferers on Earth complain of a particular set of symptoms that include chronic headaches, ringing in the ears, and diplopia (double vision). None of the ISS astronauts who suffered vision changes reported these symptoms. So researchers went back to the drawing board and tried to figure out another explanation. What about cerebrospinal fluid (CSF), they wondered? After all, any change in CSF pressure might affect the eye because an increase in CSF pressure would increase pressure on the optic nerve. CSF pressure, and several other factors that could possibly explain the vision changes, were assessed in a NASA survey that examined 300 astronauts. The results of that survey revealed that 48 percent of long duration crewmembers had experienced vision problems. The survey also found that, for some crewmembers, the vision changes continued for months or years following their return to Earth. To help the astronauts deal with the problem while researchers tried to figure out what was going on, NASA began issuing special glasses to improve visual acuity. In the meantime, NASA set up a task force to investigate the problem. Created in 2011, the VIIP task force developed a suite of experiments for ISS astronauts to participate in, one of which was the aforementioned Dilution Measures test (other tests include magnetic resonance imaging and optical coherence tomography).

While Kelly was busy collecting samples of body fluids, Kopra was beginning his Biological Rhythms experiment, a 48-hour study that investigates the effects of long-term exposure to microgravity on the cardiac system. To monitor his electrocardiographic activity, Kopra donned a Holter Monitor and an Actiwatch sleep monitor. Meanwhile, on the other side of the station, Peake was busy activating the Advanced Colloids Experiment (ACE), which required him to configure the Light Microscopy Module (LMM) to study three-dimensional structures resulting from particles suspended in a fluid. The structures, known as self-assembled colloidal structures, are valuable to engineers studying the design of advanced materials. Once he was finished tending to the ACE, Peake moved on to the Cardiac and Vessel Structure and Function with Long-Duration Space Flight and Recovery Checkout experiment (abbreviated to Vascular Echo), which required setting up the Cardiolab Portable Doppler Driver. The reason for flying the experiment on the ISS was because flight surgeons had observed stiffening of the arteries and hypertension in some returning crewmembers and it was hoped the Vascular Echo could observe changes in blood vessels during the flight that might make it easier for flight surgeons to understand the mechanism behind flight-induced hypertension. In between tending to science, the crew had its normal shopping list of maintenance tasks, which included tending to the Waste and Hygiene Compartment and myriad other jobs listed below.

Planned Activities for January 4, 2016

- Morning Inspection. Testing ПСС СМ
- Laptop RS1(2) reboot
- Inspect connectors СМ-У, РУ2, РУ4, РУ5 on МНР-НС behind panel 139 of АСУ and pipeline 5182-03 from ДКиВ to МНР in its transparent section
- Fluid Shift experiment ops
- Fine Motor Skills
- BLR48 – equipment configuration
- Video coverage for Science 2.0 TV channel
- CARDIOVECTOR. Experiment run. Tag-up with specialists
- BSA battery charging
- Preparation of items for disposal on Cygnus OA-4
- ACE experiment ops
- DOS3D – photo of DOSIS passive sensors after installation
- FIR ops
- SEISMOPROGNOZ. Info d/l from [МКСД] hard drive
- CMS – replace batteries
- EMU – reconfiguration after retrieval from transport container
- MOTOCARD. Experiment run
- VEG-01 – seedlings imagery
- HMS – inspect defibrillator
- Unstow Progress №431 (on DC1) and IMS update
- ER2 ops
- HABIT – video coverage
- WRS – recycle tank fill
- Tighten clamps of QDs between MRM2 and Soyuz №718
- JRNL – log entry
- Repair works. Installation of cover sheets on SM interior panel №334
- WHC tank and hose change-out
- EMU – reconfiguration after retrieval from transport container
- Replace [ПФ1], [ПФ2] dust filters and clean [В1], [В2] fan grilles in MRM2
- Unstow Progress №431 (on DC1) and update IMS
- VECHO
- Refill (separation) of [ЕДВ (КОВ)] for ELEKTRON or [ЕДВ-СВ]
- SOZh maintenance
- IMS update
- Update anti-virus base on [ВКС] laptops
- Daily Planning Conference (S-band)
- Preparation for anti-virus scanning on [ВКС] laptops
- Preparation of reports for Roskosmos site and social networks

- Video coverage for [ВГТРК] TV company
- ECON-M. Observation and imagery

Completed Task List Items

- EPO GV MX Kick Off (Completed on Saturday)
- EVA ECWS OBT (Completed on Saturday)
- EPO-GV RS (Completed on Sunday)
- EPO-GV-S2EC (Completed on Sunday)
- ESA-PAO-REC-DOMLIFE (Completed on Sunday)
- 45S USOS Unpack
- EVA RET inspection

Ground Activities

- Nominal system commanding
- MT Power-up and Translate (Worksite 4 – Worksite 2), SPDM Unstow

Three-Day Look Ahead

- Tuesday, 01/05: EVA loop scrub/RET inspection/EMU water conductivity measurement/DOUG review/Tool Build, MSS File Uplink for RRM TB4 return
- Wednesday, 01/06: Fluid Shifts Ultrasound, JEMAL depressurization/vent, EMU Resize, EVA ECWS OBT/Tether Inspection, Sprint setup, RRM TB4 removal
- Thursday, 01/07: Sprint VO2 ops, EVA procedures review/conference, MT Translate (WS4 – WS7)

BRITAIN'S FIRST SPACEWALK

On January 15, Peake and Kopra (Figure 6.14) headed for the airlock and began their pre-breathe, after Kelly had prepared their suits. Waiting in Mission Control was Reid Wiseman, who was working as EVA Capcom for the duration of the spacewalk. At 12:48 UTC, the two Tims kick-started U.S. EVA-35 as they exited the airlock, with Kopra leading the way as lead EVA or EV-1. As Peake clambered out of the airlock, his suit sporting the Union Jack, it marked the first time a British astronaut had ventured outside a space vehicle. Once outside the airlock, Kopra and Peake completed the checks on their suits and began making their way to the work site on the Starboard Truss Segment (S6), which is where the balky SSU was. Once there, they set up their work site and got to work.

The core of the spacewalk was a 31-minute orbital 'night', which mission planners had deemed long enough to remove the old SSU and install the replacement that had been in the box of spare parts since 1999. The SSU weighs about 100 kilograms and is located at the base of the Mast Canister Assembly of the array wing. The first item on the work agenda was breaking out the ratchet wrench, which Kopra used to break the torque on the bolt holding the SSU in place. Meanwhile Peake released the straps holding the spare SSU in its bag and made sure he had his Pistol Grip Tool ready. Once Mission Control confirmed the solar array was not generating power, Kopra and Peake made quick work of the task, removing the failed SSU by releasing the bolt and sliding the SSU out of its location. Once Peake (Figure 6.15) had inspected the failed SSU and confirmed that the fault did

Figure 6.14. Tim Peake and Tim Kopra prepare for their EVA. Credit: NASA

indeed lie within the unit, the SSU was stowed and Peake handed Kopra the spare unit, which he slid into position. Using the wrench, Kopra tightened the bolt and the job was done with time to spare.

As the two spacewalking astronauts were preparing to attend to other tasks, an off-nominal carbon dioxide reading in Kopra's suit indicated a potential problem. One of the problems with the suits is the occasional failure of the carbon dioxide sensors when they come into contact with sweat. Mission Control reviewed the data and instructed Kopra to go ahead with the rest of the spacewalk. As Kopra and Peake went about packing up the failed SSU and taking engineering photos, the Sun rose above the horizon and the two made tracks to complete the remaining tasks on the task-list. For Kopra, this meant heading towards the Z1 Truss and then to Node 3 to install the Non-Propulsive Vent (NPV), a unit that releases atmospheric waste gases overboard. After that was installed, Kopra headed for the Pressurized Mating Adapter-3 (PMA-3), where he removed a launch restraint in preparation for a future spacewalk.

Peake, meanwhile, was engaged in another maintenance task on the nadir-facing side of the Destiny module, where he deployed an International Docking Adapter (IDA) cable reel. Next on Peake's list of things to do was to connect a cable to the Multipurpose Laboratory Module. He was half way through this task when Kopra, who had just completed his NPV work, reported water accumulating inside his helmet. The water leak wasn't as serious as the intrusion suffered by Luca Parmitano three years earlier (see sidebar), but Mission Control, mindful of lessons learned in that event, terminated the EVA[2]. The globule of water

---

[2] An *EVA Termination* is different than an *EVA Abort*. An EVA Termination permits the crew to make safe the worksites and return to the airlock to complete a normal repressurization, whereas an EVA Abort requires an immediate return to the airlock to complete an emergency repressurization.

Figure 6.15.   Tim Peake during his EVA, January 15, 2016. Credit: NASA

inside Kopra's helmet was about the size of a golf ball at the time he and Peake started heading for home, but the accumulation had increased when the two spacewalkers arrived back at the airlock. First into the airlock was Kopra, who quickly moved into the Equipment Lock so he could remove his helmet as soon as possible. He was soon joined by Peake and the two completed a normal depressurization. The EVA – the 192nd in support of ISS operations – had lasted four hours and 43 minutes.

When spacesuits fail

In July 2013, Expedition 36 astronauts Luca Parmitano and Chris Cassidy were conducting EVA-23, which involved various cable routing and equipment reconfiguration tasks. All was proceeding well until 44 minutes into the spacewalk, when Parmitano reported water at the back of his helmet. Mission controllers thought it might a drink bag leaking but after some troubleshooting it became clear water was leaking from elsewhere and the EVA was terminated.

> "The water has also almost completely covered the front of my visor, sticking to it and obscuring my vision. At that moment, as I turn 'upside-down', two things happen: The Sun sets, and my ability to see – already compromised by the water – completely vanishes, making my eyes useless; but worse than that, the water covers my nose – a really awful sensation that I make worse by my vain attempts to move the water by shaking my head. By now, the upper part of the helmet is full of water and I can't even be sure that the next time I breathe I will fill my lungs with air and not liquid."

Excerpt from Parmitano's blog

As Parmitano made tracks for the safety of the airlock, his communications shorted and the volume of water filled his helmet to such an extent he could barely see. Analysis revealed the amount of water in Parmitano's helmet to be almost 1.5 liters. The incident prompted setting up a Mishap Investigation Board, which revealed that the Fan Pump Separator (FPS – the component that circulates air and oxygen through the Oxygen Ventilation Circuit) had failed. The separator was removed from Parmitano's suit (EMU-3011) and a replacement was ferried up to the ISS on the Progress M-20M mission. Closer inspection of the failed unit suggested the cause of the failure was contamination inside the FPS. Ultimately, the Mishap Investigation Board made 49 recommendations based on their analysis of the event, 16 of which had to be completed before spacewalks could continue.

With the airlock pressurized, Kopra moved to the Equipment Lock with the assistance of Malenchenko and Volkov. What was troubling was the fact that the suit – EMU #3011 – was the same one that had been worn by Parmitano during his 2013 EVA.

Table 6.4 Standard crew day[A]

06:00 AM - Crew wake. Clean up, eat, read news/messages uplinked overnight.
07:30 AM - Morning Daily Planning Conference (DPC). Astronauts/cosmonauts sync with Houston, Huntsville, Munich, Tskuba, and Moscow before executing their day.
07:55 AM - Work prep, review procedures and gather stowage to support the day's activities.
08:15 AM - Crew available work time. Science experiments, preventative and corrective maintenance, visiting vehicle preparations, stowage operations, environment sampling (acoustics, surfaces, water), public affairs events, miscellaneous medical tasks including daily 2.5 hrs. of exercise.
01:00 PM - Lunch
02:00 PM - Crew available work time (see above).
06:15 PM - Evening work prep, review procedures and timeline for next day.
07:05 PM - Evening DPC. Discuss comments/questions about the day executed. Brief highlight for changes to tomorrow's plan (if required).
07:30 PM - Dinner, relax, email, downlink images, watch a movie, look at Earth!
09:30 PM - Crew Sleep (8.5 hrs.)

[A] Source: NASA Flight Operations Directorate

Back to the routine

With the first spacewalk of 2016 out of the way, the crew settled into their daily routine. Table 6.4 gives you an idea of what Peake's typical workday involved, but in reality, a routine day on a vehicle as complex as the ISS is something of a rarity. Systems fail, procedures are changed, timelines reconfigured, and the plan needs to be amended. And that plan, like just about everything at NASA, has an acronym – the OSTPV, for Onboard Short Term Plan Viewer. The OSTPV is a timeline that shows the ISS schedule, and is divided into bands for the crew and into bars for actual activities. The plan generally spans a period of a few days to up to two weeks and is developed using NASA's Consolidated Planning System (CPS). If schedules need to be changed, then flight controllers step in and manipulate it accordingly.

A second spacewalk

After the 192nd EVA, the next major event on ISS was the 193rd. On February 4, 2016, Malenchenko and Volkov opened the hatch of the Pirs airlock and steadily worked their way through the 'task jar' for the day. This included swapping out materials science space exposure samples, installing handrails for future EVAs, and testing film coatings that may one day be used to repair damaged hull panels. At the close of the four-hour 45-minute spacewalk, Malenchenko had clocked up his sixth EVA and Volkov his fourth. The ISS had now racked up a total of 1,204 hours of EVA, or more than 50 days.

With two EVAs out of the way, the crew focused on the next job in NASA's task jar – the undocking of the Cygnus spacecraft, which had arrived on station on December 9, 2015 on a United Launch Alliance (ULA) Atlas V rocket. The mission had marked ULA's fourth contracted commercial resupply mission. On February 19, Kelly and Kopra commanded the robotic arm to release the Cygnus and, once the spacecraft was a safe distance from the station, its engines fired, setting the vehicle on course for a deorbit burn and re-entry.

Figure 6.16.   Yes, astronauts have a sense of humor! Thinking that he might need cheering up as he approached the end of almost 12 months in space, Mark Kelly sent his brother Scott a gorilla suit for his 52nd birthday. A word about the cost of sending that suit into orbit: The Soyuz FG costs around $50 million per launch and can carry 16,350 lbs of supplies to the station. This means it costs around $3,059 for every pound sent to the ISS. The gorilla suit, by California Costumes, weighed around 4.3 lbs and cost around $52. Which means the cost of getting that suit to the ISS was about $13,200. Kelly wasn't the only astronaut who had a special care package delivered: Peake was sent a tuxedo t-shirt at a cost estimated to be around $1,600. Image taken from video. Credit: NASA

A bit of fun

Spending the best part of a year on orbit, cooped up in a space not much larger than your average three-bedroom house, can make even the most level-headed astronaut feel the need to ... well, loosen up. Which explains the antics in Figure 6.16. Sort of. The suit was a gift from Scott Kelly's brother Mark and it arrived on station in a care package. The video of the on-orbit antics, accompanied by the song '*Yakety Sax*' (the British '*Benny Hill*' theme tune), shows Kelly bursting out of a box wearing said costume and proceeding to chase Tim Peake through the station.

Scott Kelly
@StationCDRKelly
Needed a little humor to lighten up a #YearInSpace. Go big, or go home. I think I'll do both. #SpaceApe
4:13 PM - 23 Feb 2016

The 'goin' ape' in space was typical of the sort of antics ISS followers had come to expect from one of NASA's longest duration astronauts. During the course of Kelly's near year-long stay on the station, Twitter fans had been treated to viewing the Super Bowl from orbit, a discussion on the dangers of space pee and an introduction to the challenges of playing ping-pong in space. Not to be outdone, Peake sent a video message the following day to the Brit Awards and congratulated Adele for her four awards.

Veteran Station Crew Returns to Earth after Historic Mission

That was typical of the taglines in the space media on March 1, 2016, a date which heralded the return of Kelly and Kornienko from their epic 340-day, 143-million-mile stay on the ISS. With the landing of the Soyuz TMA-18M spacecraft in Kazakhstan, Kelly had racked up 520 days in space, with Kornienko not far behind on 516. Volkov, meanwhile, arrived back on Earth with a total of 548 days logged in space, taking the family duration record from his father. The landing of Expedition 46 meant the launch of Expedition 47 was only days away, which is where we pick up the narrative in Chapter 7.

7

Principia Part 2

Figure 7.0. Credit: ESA

© Springer International Publishing AG 2017
E. Seedhouse, *Tim Peake and Britain's Road To Space*, Springer Praxis Books,
DOI 10.1007/978-3-319-57907-8_7

NEW ARRIVALS

With the departure of Expedition 46 on March 2, the ISS complement was reduced to three. But not for long. On March 19, Soyuz TMA-20M docked with the orbiting outpost, carrying Expedition 47 crewmembers Aleksei Ovchinin, Oleg Skripochka and Jeffrey Williams (see Appendix IV). Once again, it was a crew with a heavy military background, with Ovchinin a retired air force officer and Williams a retired army officer. Saying goodbye to old crewmates and saying hello to new ones is one of the many challenges of the long duration spaceflight game. Whereas on a nuclear submarine you know you're stuck with the same faces for months on end, on an orbiting outpost those faces change, and that means you have to go through the process of getting to know one another all over again. Not that big a deal, but a little disruptive all the same, no matter how well trained you are.

Aleksei Ovchinin

The Commander of Expedition 47 was Aleksei Ovchinin (Figure 7.1), a retired Russian Air Force colonel. After a spell at the Borisoglebsk Higher Military Pilot School, Ovchinin transferred to the Yeisk Pilot School, graduating as a pilot-engineer in 1992. He spent six

Figure 7.1. Aleksei Ovchinin. Credit: Russian Space Agency

years working as a pilot instructor at Yeisk, before moving on to Krasnodar Military Aviation Institute where he held the position of Commander of the aviation section until 2003. During the course of his military career, Ovchinin logged more than 1,300 hours flight time, which made him good cosmonaut material. He was selected as a Cosmonaut Candidate in 2006, and followed the standard two-year training and evaluation program before being awarded his qualifications as Test Cosmonaut in June 2009. After spending some time training on the ISS Mini Research Modules, Ovchinin was finally certified as a cosmonaut in April 2010 and looked forward to his first mission assignment.

His first opportunity looked like it might be Soyuz TMA-16M in 2015, but after the one-year mission was confirmed, Ovchinin found himself on the waiting list, occupying his time with cosmonaut-related activities such as caving in Sardinia. After serving as back-up to Soyuz TMA-16M, Ovchinin was bumped up to prime for the 47/48 crew. As part of his preparations he spent time completing winter survival training, before studying for his ISS and Soyuz TMA exams in March 2015. With those out of the way, Ovchinin was given the green light by the Russian State Commission as back-up Commander for the TMA-16M crew. Once the TMA-16M crew had launched, Ovchinin's training switched to prime for the TMA-20M flight.

Oleg Skripochka

Flying with Ovchinin was Oleg Skripochka (Figure 7.2). A pilot-cosmonaut with a background in mechanical engineering (with a specialty in rocket construction), Skripochka had previously flown as part of the Expedition 25/26 crews, during which he was lucky enough to have conducted three spacewalks. Before becoming a cosmonaut, Skripochka

Figure 7.2. Oleg Skripochka. Credit: Russian Space Agency

had worked as a technician for Energia, Russia's rocket manufacturer, where he specialized in developing ground support systems for cargo spacecraft such as the Progress. Skripochka's cosmonaut career started when he was selected as a cosmonaut candidate in 1997. His two-year evaluation period started in January 1998, and he received his qualification as Test Cosmonaut in December 1999. After qualifying as a cosmonaut, Skripochka was assigned as a back-up to the sixth ISS visiting crew, but this mission was cancelled in the wake of the *Columbia* tragedy when subsequent crew transfer missions were restructured. His next flight assignment was as back-up to Expedition 19 in 2006, a role that was followed by another back-up assignment to the ISS-17 mission. Skripochka finally received a prime crew flight assignment in 2008 when he was confirmed for Expedition 25/26. Skripochka and his crewmates, Aleksandr Kaleri and Scott Kelly, were the first to be trained on the upgraded Soyuz TMA-M variant that featured an enhanced digital flight control system and more up-to-date flight computers. After the standard two years mission training, Skripochka, Kaleri and Kelly served as back-up crew to the Soyuz TMA-18 mission in April 2010, before being confirmed as prime crew for their October 2010 flight.

After blasting off on October 7, Skripochka's Soyuz TMA-01M vehicle performed its routine 34 orbits of the Earth before docking with the ISS on October 10. Waiting to greet Skripochka's crew were Expedition 25 crewmembers Doug Wheelock, Shannon Walker and Fyodor Yurchikhin. A month later, Skripochka and Yurchikhin stepped outside the station to conduct a six-and-a-half-hour maintenance spacewalk. Shortly after the spacewalk, the Expedition 26 crew waved goodbye to the Expedition 25 crew and worked as a crew of three for two weeks until November 20, when they were joined by the Expedition 27 crew of Dimitri Kondratyev, Cady Coleman and Paolo Nespoli. In January 2011, Skripochka had the opportunity to conduct the second of his three spacewalks when he joined Kondratyev on a five-and-a-half-hour excursion to install communications equipment. Three weeks later, the two spacewalkers were outside again to install radio antennae and deploy a satellite. Skripochka's third spacewalk brought his total time outside the ISS to almost 17 hours.

Eight days after Skripochka's third spacewalk, the Expedition 26 crew welcomed the Shuttle *Discovery*, which delivered the Permanent Multipurpose Module, the ExPRESS Logistics Carrier, and Robonaut. Shortly after Discovery's departure, it was Skripochka's turn to leave the station. On March 16, Soyuz TMA-01M undocked from the station and headed home to the snowy Kazakh landscape after a flight of 159 days, 8 hours and 43 minutes. On returning to Earth, Skripochka slotted into regular ISS training while waiting for his next mission assignment, which was confirmed in January 2014 when he was nominated as prime for Expedition 47/48 and as back-up to Expedition 45/46.

Jeffrey Williams

The third member of the Expedition 47/48 crew was NASA astronaut Jeffrey Williams (Figure 7.3), a retired Army officer and veteran of three spaceflights. The oldest of the new crew, Williams was making his fourth spaceflight at the age of 58. Prior to service as an astronaut, he had worked as an aviator in military service in the 1980s, before being selected for a military assignment at the Johnson Space Center (JSC) where he worked on Shuttle launch and landing operations, and as a pilot in the Shuttle Avionics Integration Laboratory

Figure 7.3. Jeffrey Williams. Credit: NASA

(SAIL). Not a bad job to put on your astronaut resume I guess! After five years working at JSC, Williams enrolled at the Naval Test Pilot School, where he added more flying hours to his logbook (during his 27 years in the military, Williams logged more than 3,000 hours across 50 aircraft types). He was selected as an astronaut in 1996 and completed the requisite two years of training before being assigned to duties within the Astronaut Office, specializing in ISS programs, extravehicular activity, and developing upgrades to the Shuttle cockpit. In November 1998, Williams received his first flight assignment as a Mission Specialist on board the STS-101 (*Atlantis*) ISS assembly mission. Following 18 months of training, the STS-101 mission lifted off on May 19, 2000. At this point in the ISS program, the orbiting outpost consisted of just Zarya, the Node 1 Modules and the Pressurized Mating Adapter, which meant there was plenty to keep the *Atlantis* crew busy. While docked with the embryonic station, the crew transferred several tons of equipment from the Shuttle to the ISS. This equipment included light fixtures, air ducts, radio equipment and fire extinguishers. Before heading back to Earth, Williams completed a six-hour 44-minute spacewalk, together with James Voss, to fit components to the station exterior.

On his return to Earth, Williams served as a back-up crewmember to Expedition 12 before slotting into a prime crew assignment for Expedition 13, the last of the two-person 'caretaker' crews. As with so many ISS-bound astronauts, Williams participated in a NEEMO mission on board Aquarius. Williams' mode of transport to ISS the second time around was a little different than his *Atlantis* experience. Instead of being ferried up by one of the Shuttles, which had not yet been cleared to resume full ISS operations after

Columbia, Williams' second spaceflight was on board the Soyuz. Soyuz TMA-8 lifted off on March 30, 2006, carrying Williams, Commander Pavel Vinogradov and Brazilian spaceflight participant Marcos Pontes. Once Pontes had departed the station a week later, it was just Williams and Vinogradov on orbit.

For the next three months, the pair tended to various tasks, including a busy science schedule, receiving cargo ferried up by the Progress M-56 vehicle, and conducting a spacewalk wearing the Russian Orlan suit to fix the Russian Elektron oxygen generation system. After their lonely stint on orbit, Williams and Vinogradov were keen to welcome the Shuttle *Discovery* (STS-121) in July. In addition to ferrying ESA astronaut Thomas Reiter to the station to bring the crew complement back up to three, *Discovery* also delivered a shopping list of cargo using the Multi-Purpose Logistics Module. After an eight-day stay, *Discovery* departed the ISS, and the station's crew of three spent the next two months performing standard housekeeping and science tasks before preparing for the next Shuttle visit, this time by *Atlantis* on STS-115 in September. During its six-day stay, STS-115 resumed station construction operations, and included three spacewalks to install the P3/P4 Truss segment and a new solar array wing. After handing over the station to Expedition 14, Williams and Vinogradov took their places in *Atlantis* for the return ride. The landing, on September 29, 2006, marked the end of a mission that had lasted 182 days, 22 hours and 43 minutes.

On returning to JSC, Williams entered into the familiar routine of ISS support operations before being assigned as back-up to the Expedition 19 crew. This required him to gain the qualifications to fly on the Soyuz as a Flight Engineer. Two years after his return from the ISS, Williams was assigned as a prime crewmember to Expedition 21 and began the one-year training preparation, together with Commander Maksim Suraev and Canadian spaceflight participant Guy Laliberté of Cirque du Soleil fame. The trio launched on Soyuz TMA-16 on September 30, 2009. Two days later, the Soyuz docked with the ISS and Williams, Suraev, and Laliberté joined crewmembers Frank DeWinne of ESA, Russian cosmonaut Roman Romanenko, CSA astronaut Robert Thirsk, and NASA astronaut Nicole Stott. A few weeks later, the ISS was visited by *Atlantis* on Shuttle mission STS-129. In addition to delivering the Express Logistics Carriers 1 and 2, *Atlantis* also performed the Shuttle program's final crew rotation when it returned Stott to Earth on November 23.

Expedition 21 continued a little while longer though, with DeWinne, Romanenko and Thirsk staying on board until December 1, before returning to Earth on board Soyuz TMA-15. With Expedition 21 departed, the station was once again occupied by just two crewmembers: Williams and Suraev (this also marked the last time the ISS was manned by just two crew). Unlike the last time, when Williams shared the orbiting outpost with just one crewmember for three months, this two-man increment lasted just one week thanks to the launch of Soyuz TMA-17 on December 21, which delivered the Expedition 23 crew of Oleg Kotov, Soichi Noguchi and Timothy Creamer to the ISS. Following a visit by *Endeavour* (STS-130) in February 2010, Williams and Suraev prepared for their return the following month via the Soyuz, which landed on March 18, 2010. For Williams, it was another long duration mission – 169 days – to chalk up in his logbook. For most astronauts this would be enough, but Williams wasn't finished yet. Next up was an assignment as back-up to Scott Kelly's one-year mission (that spanned Expeditions 43 through 46) before being nominated as a prime crew to Expedition 47/48 together with Ovchinin and Skripochka.

BEAM

Not long after the Expedition 47/48 crew arrived, Orbital ATK's Cygnus cargo vehicle launched from Space Launch Complex 41 in Florida, carrying 3,400 kilograms of crew supplies and hardware. Included in the cargo manifest was a 3-D printer and a variety of science experiments to keep the crew busy. Cygnus arrived at the station on March 26, where it was grappled by the station's robotic arm under the guidance of Tim Peake and Timothy Kopra (Figure 7.4). More vehicular activity took place four days later, when the Progress M-29M vehicle, fully laden with trash, undocked from the Zvezda Service Module. A day later, the Progress MS-02 cargo vehicle launched from the Baikonur Cosmodrome, loaded with 2,400 kilograms of food, fuel and supplies. The new Progress arrived at the station on April 2, by which time the crew were no doubt well-practiced in the routine of offloading supplies. The flurry of Cygnus and Progress activity was pretty much business as usual for the crew, but the next launch proved a little different. That's because SpaceX's Dragon capsule carried a very unique cargo. Launched on April 8, 2016, on a Falcon 9 rocket from Space Launch Complex 40, SpaceX's CRS-8 manifest included Bigelow Aerospace's expandable habitat – the Bigelow Expandable Activity Module or BEAM (see sidebar, and Figures 7.5 a, and b).

Figure 7.4. Tim Peake and Tim Kopra manning the robotic workstation in the Cupola while berthing the Cygnus. Credit: NASA

Figure 7.5(a). BEAM concept art. BEAM was attached to the Tranquility module using the Canadarm2, before being filled with air in preparation for a two-year technology test period, during which astronauts will conduct tests to validate the performance and capabilities of expandable habitats. At frequent intervals, crewmembers will enter the BEAM to take measurements and monitor the habitat's performance, to help inform engineers about designs for future habitats. Learning how such a habitat performs in the thermal environment of space – and how its materials react to radiation, micrometeoroids and orbital debris – will provide key data to engineers, who may need to make some tweaks to the design before a fully-fledged BEAM station can be flown. After the two-year test, BEAM will be detached from the ISS to burn up in the Earth's atmosphere. If BEAM performs to expectations, it could lead to future development of larger and more permanent expandable habitation structures in LEO, and for crews traveling beyond LEO. The BEAM project is co-sponsored by NASA's Advanced Exploration Systems (AES) Division and Bigelow Aerospace. Credit: NASA

"The world of low Earth orbit belongs to industry. You need to understand where we're going. A Bigelow module may be the next thing that begins to replace some of the functions of the International Space Station. Low Earth orbit infrastructure belongs to industry ... If we don't have a viable, vibrant low Earth orbit infrastructure supported by them [commercial industry], we're not getting there [Mars]."

Charles Bolden, NASA Administrator, January 26, 2015

Figure 7.5b. BEAM concept art, showing the expandable habitat at full inflation. Credit: NASA

BEAM

Inflatable habitats are nothing new. In fact, the idea has been around since the 1960s, although it wasn't until the 1990s that NASA developed the concept, which became known as TransHab. Like so many good ideas, TransHab was cancelled by Congress, but space mogul Robert Bigelow recognized the potential of the idea and bought the rights to the patents to develop his own version of inflatable space habitats. He flew two unmanned versions – Genesis I and II – in 2006 and 2007. They're still up there. But Bigelow's vision was for manned inflatable space stations and he needed some way of testing the concept on orbit, which is how the Bigelow Expandable Activity Module – BEAM – came to be. A technology demonstrator, the BEAM was flown as part of a $17.8 million NASA Advanced Exploration Systems (AES) Program contract. After being delivered to the ISS, the BEAM was expanded on May 28, 2016, over the course of more than six hours. During its two-year tenure on the station, the BEAM will test and validate expandable habitat technology by:

  i. Demonstrating launch and deployment of an expandable module
 ii. Determining the radiation protection of such a structure
iii. Demonstrating the thermal, structural and mechanical design performance

BEAM by the numbers (deployed dimensions):

- Length: 4.01 meters
- Diameter: 3.23 meters
- Mass: 1,413 kilograms
- Volume: 16 m³
- Radiation shielding: Multiple layers of closed-cell vinyl polymer foam
- Impact protection: Whipple Shield

"Well, we're excited to add another – what is essentially – vehicle on to the Station. This is another example of where the commercial industry has been innovative in their techniques. BEAM will be a great way to test out the thermal characteristics of this new type of module, along with its radiation protection. It's going to be a neat thing. I think any time we can bring up new modules to Space Station we're going to be excited about that. We have a big Station here but there's room to make it bigger. So, if we can add additional modules to test out the future of space exploration – and the way humans are going to interact with modules – I think it's a great idea."

Tim Kopra

"The International Space Station is a uniquely suited test bed to demonstrate innovative exploration technologies like the BEAM. As we venture deeper into space on the path to Mars, habitats that allow for long-duration stays in space will be critical capability. Using the station's resources, we'll learn how humans can work effectively with this technology in space, as we continue to advance our understanding in all aspects for long-duration spaceflight aboard the orbiting laboratory."

*William Gerstenmaier, NASA Associate Administrator
for Human Exploration and Operations, speaking at
NASA Headquarters in Washington on January 16, 2013*

The potential for BEAM

Bigelow's BEAM (Figure 7.6) could be a breakthrough technology that recalibrates how astronauts live and work in space. Not only are expandable structures much cheaper to launch and deploy, this new type of habitat dramatically increases the living space available to crews and the structure itself provides greater radiation protection than traditional aluminum constructions. Composed of super tough Vectran and Kevlar, Bigelow's expandable modules also provide much better protection against orbital debris than the ISS modules. While the current BEAM is smaller than the ISS modules, Bigelow[1] is thinking big and has plans to orbit a BA-330, a full-scale habitation module large enough to accommodate a crew

---

[1] If you are interested in the Bigelow story, the go-to reference publication is *Bigelow Aerospace*, written by the author, and published by Springer-Praxis.

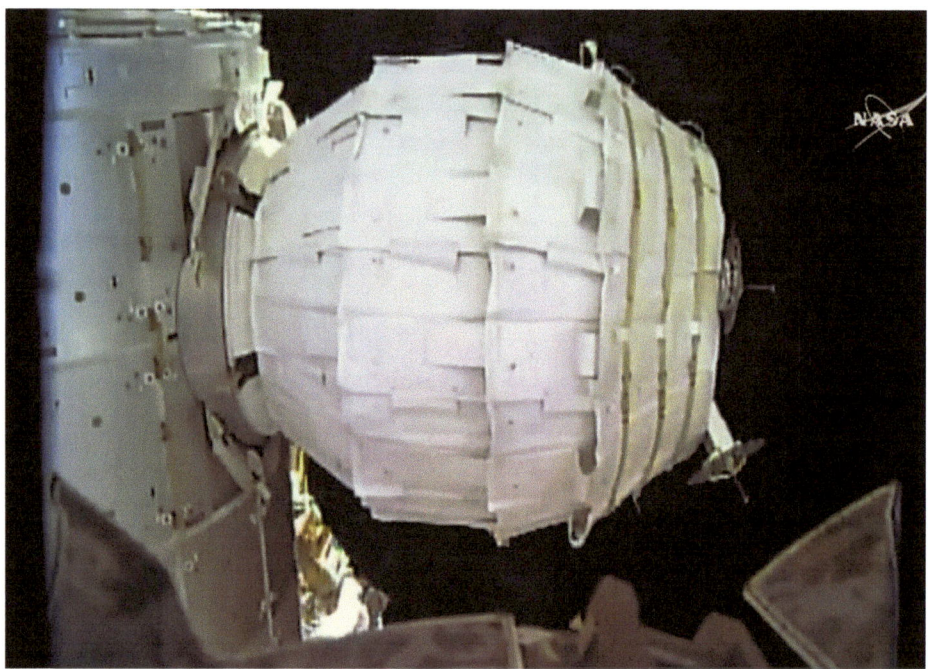

Figure 7.6. BEAM: actual, expanded. Credit: NASA

of six. The BA-330 may also serve as the basis for a modular commercial space station for sovereign customers who have no space program of their own (see sidebar). The BA-330 has a pressurized volume of 330 cubic meters and measures 6.7 meters in diameter and 13.7 meters in length. Dwarfing even the capacious BA-330 is the BA-2100 behemoth, an expandable with more than twice the pressurized volume of the entire ISS.

"The 'Bigelow Expandable Activity Module', or the BEAM, is an expandable habitat that will be used to investigate technology and understand the potential benefits of such habitats for human missions to deep space. The ISS is an excellent platform to test and demonstrate explorations systems such as the BEAM. And NASA expects that the BEAM project will gather critical data related to structural, thermal, and acoustic performance, as well as radiation and micrometeoroid protection. All of these data are essential to understanding the technology for future astronaut habitats for use in long-duration space travel."

Charlie Bolden

Future (expandable) space stations

All sorts of suggestions for commercial space stations have been made over the years. Bigelow's suggestions include using the modules as medical facilities, resupply depots and deep space exploration habitats. They could also be used as manufacturing facilities for products that can only be made in microgravity, or they could be transit ports for spacecraft heading beyond Earth orbit. They could be used as test platforms, as research facilities, or as a biological containment facility. How they will be used is still the subject of debate, but when the ISS is de-orbited and splashes into the Pacific sometime in 2028, there will be no orbiting destination to go to, which should open up a market for one or more of Bigelow's commercial stations. What kind of station will be the most commercially successful? It's difficult to say, but commercial crews and cargo will need thriving and growing destinations to have any significant growth potential and to make a return on investment. Space tourism would be greatly helped with a destination that has room for people to move and float around in, but that's just one part of the equation.

So, if you happen to be a potential Bigelow sovereign customer, what will you get for your $51.25 million ($26.25 million for the launch on board the Dragon plus $25 million for your 60-day stay on board a BA-330)? Well, that price includes the costs of the launch and range fees and the cost of recovery, in addition to the cost of training the client. But will this business of launching customers to inflatable habitats be profitable? That's hard to say, because there are so many variables. For one thing, launch costs will be a function of traffic and the frequency and quantity of flights. Launch costs can also be a function of the ability of the launch provider to become more cost efficient in the fabrication of their vehicle (the launch vehicle and/or the capsule). In setting out his business plan, Robert Bigelow is always keen to emphasize that his concept of space outpost utilization is not about space hotels, despite what the media say. Instead, Bigelow imagines building commercial space habitats that can accommodate a variety of functions: from governments aiming to pursue their own space programs to researchers wanting to conduct science. Ultimately, Bigelow feels that it will be applications derived from space research that will drive the engine of space development.

The business case

When commercial real estate developers build an office building, they usually don't have to pre-sell all the space before they start construction. Instead, the company usually signs an anchor tenant to secure financing. Could such a model work for commercial space stations? Obviously, it would help to have government and commercial tenants signed to get financing for a commercial station, which is why Bigelow has agreements in place with more than half-a-dozen sovereign nations. But will they take the risk? Well, no-one really knows if revenue can be guaranteed, but once it *does* happen, this commercial space station business suddenly becomes more interesting to the financial community. The question then becomes: who will the anchor tenant be? The answer, of course, is the government, simply because no one else has deep enough pockets. Whether that happens is in the hands of the politicians and NASA's long-term vision, which is something Congress still can't agree on. NASA wants to move beyond LEO, and turning LEO over to potentially more

Table 7.1 Bigelow's Business Case in a Nutshell

Capital costs	
A. Cost of building BA-330	$125 million
B. Cost of launching BA-330 (via Falcon Heavy)	$80 million
C. Capital cost (A + B)	**$205 million**
Operating costs	
D. Three cargo resupply flights/$100million per flight	$300 million
E. Operation Support costs	$50 million
F. Annual costs (D + E)	**$350 million**
Revenue	
Monthly Revenue (assuming maximum occupancy)	
G. 3 x $25 million	$75 million
Annual Revenue	
H. (G x 6)	**$450 million**
Return on Investment (assuming 10-year lifespan)	
I. Annual operating costs (10% of C + F)	$320.5 million
J. 10-year annual revenue (H x 10)	$4.5 billion
K. 10-year profit (J - 10 x I)	$1.295 billion

cost-effective business models such as orbiting BA-330s and BA-2100s would free up resources to develop payloads for exploration-class missions. But Bigelow has cautioned that his company cannot fund the entire operation without some commercial partners and they need to sign enough customers to pay for half the cost of launching the modules into LEO. Ultimately, Bigelow has acknowledged that the commercial sector will need NASA as an anchor tenant to have a business case for exploration beyond LEO. The reason? The commercial space industry is just finding its legs, and without an anchor tenant it just doesn't have the strength to execute deep space missions.

In 2017, Bigelow's BA-330 is essentially ready to launch when commercial crew vehicles become operational. Demand? Well that depends on the business case. Which foreign and domestic companies could benefit from utilizing expandable, orbital space habitats and what are the profit margins for Bigelow? Well, let's crunch some numbers (Table 7.1). If building a BA-330 costs $125 million and launching it on one of SpaceX's Falcon Heavys costs another $80 million, then each module will have a capital cost of $205 million. Now, if all three of the sections are rented for $25 million each for 60 days, the maximum revenue for a BA-330 module would be about $450 million per year. Let's assume three cargo flights are needed to resupply the module, at about $100 million per flight, and operation support costs another $50 million per module per year. That brings the annual costs to $350 million. If the module has a 10-year lifespan (Bigelow reckons the modules could last 20 years, but let's be conservative) then the total available for profit after paying off the capital investment is ~$900M per module, on just a $205M investment. That's a healthy return!

These calculations assume an occupancy rate of 100 percent, but how likely is that scenario? Truth is, we just don't know, but even with a reduced occupancy rate, the numbers look good. $51.25 million for 60 days (the cost of the flight and the occupancy) is relatively economical when you consider that spaceflight participants have paid $35 million or more for just ten days on board the ISS. And don't forget that these spaceflight participants had to spend at least six months in Star City training for the mission, as well as having to learn Russian. Spending that length of time training will no doubt prove off-putting to most of the people with the financial wherewithal to afford such a flight in the first place.

MARATHON MAN

After the BEAM had been secured to the ISS, Tim Peake turned his attention to the Principia experiments, but before we delve into the details it is worth mentioning another Brit that was working alongside him. Peake's other 'colleague' was a computer with the moniker Raspberry Pi. The Raspberry Pi was a core element of the Astro Pi mission that was designed to measure the environment of the ISS, thanks to a sensor suite capable of taking readings of temperature, pressure, humidity and the Earth's magnetic field. While the computer's suite of sensors was nothing new, what made the Raspberry Pi unique was its use as a platform for British schoolchildren to run their own code. The idea was the initiative of the UK Space Agency (UKSA), which was investigating ways to encourage children to think about applications on board the orbiting outpost. A meeting with Raspberry Pi's CEO, Eben Upton, followed and from there momentum built steadily. The next step was challenging schools to devise computer science-based experiments that Peake could operate while on orbit. Having the opportunity of flying their code in space inspired schoolchildren to submit an impressive quality of entries. One of the winning entries (there were seven) measured humidity fluctuations based on the presence of a crewmember: if a fluctuation was detected, the Pi delivered a message on its screen and a photo was taken of the crew. Another entry, dubbed Watchdog and devised by Kieran Wand at Cottenham Village College, used the Pi sensors to measure temperature, pressure and humidity, and to raise the alarm if they moved outside thresholds. Peake's role in the Pi suite of experiments was to check their progress using an app called the Master Control Program.

British Astronaut Tim Peake Sets Off-World Record Running Marathon in Space

That was the space.com headline on April 25, 2016, that alluded to the previous day's efforts of Peake while strapped to the station's treadmill (Figure 7.7). The London Marathon had taken place the previous day, and Peake had started his effort at the same time as tens of thousands of runners set out in the capital city 400 kilometers below (Peake was somewhere over the Pacific when the 10 am starting gun fired). Dressed in a red vest and black shorts, a British flag draped in the background, Peake ran the distance in a *very* respectable three hours, 35 minutes and 21 seconds. In ESA's Astronaut Training Center, monitors carried live footage of Peake running. Cheers and applause broke out among the staff once the Brit finished, in what was later confirmed as a Guinness World Record by an official from the esteemed publication who had been watching along with the ESA staff. While the thousands who pounded the streets of London had the city sights to distract them from the pain that comes with racing 26.2 miles, Peake completed his run in a windowless module just across from one of the station's three washrooms. Starting his run at a steady 7.5 mph, Peake ran a negative split by cranking up the pace to 9 mph over the final six miles.

Figure 7.7. Tim Peake, running the London Marathon, April 24, 2016. Credit: NASA

For those unfamiliar with the challenges of keeping fit in space, it is worth bearing in mind that Peake's marathon came four months into his mission. By this time, no matter how fanatically you follow the exercise plan, muscles will have atrophied, the heart will have lost mass, and the whole body will have weakened due to the deleterious effects of microgravity. Added to that huge disadvantage was the fact that Peake was loaded with only 70 percent of normal Earth bodyweight. And for those of you familiar with the pain that comes with running a marathon, consider the system Peake had to use. Take a look at the image of Peake on the treadmill. See those bungees and the harness system? They dig in, so that after just half a mile, abrasions and pressure sores start to build. Just running a mile using this system is uncomfortable. Running 26.2 miles? At the speed that Peake ran? That's a true pain fest. And you were worried about blisters? It helped that Peake has been a lifelong runner, although he prefers cross-country to marathons, and, like almost every (sane) runner, favors just about anything over running on a treadmill (during his ISS marathon, Peake used an iPad app called RunSocial, which allowed terrestrial runners to follow him while he ran).

"Running the fastest marathon in space on the only day off from his grueling schedule is a fantastic accomplishment. Tim is a true inspiration and someone we can all look up to. Literally."

Marco Frigatti, head of records for Guinness World Records.
Frigatti adjudicated Peake's marathon via live video
from ESA's European Astronaut Centre in Cologne.

While Peake became the first *man* to run a marathon in space, he wasn't the first astronaut to do so. That honor goes to NASA astronaut Sunita Williams, who ran the Boston Marathon in 2007. Her finishing time was four hours and 24 minutes, although her time can't be compared to Peake's time because different treadmills were used. The equipment Williams used was the Treadmill with Vibration Isolation and Stabilization System (TVIS), which counteracted the vibrations created by running on the system by using a gyroscope. In contrast, the COLBERT (Combined Operational Load Bearing External Resistance Treadmill) system that Peake ran on doesn't use a spinning gyro and therefore provides a more stable running surface.

Inflating the BEAM

Next on the space station calendar was the departure of SpaceX's Dragon. The vehicle, which had been attached to the station since April 10, was undocked from Harmony's nadir port on May 11 using the robotic arm, which was operated by ground controllers with Peake executing the command for the spacecraft's release. In what was now a familiar sequence of events, the Dragon pulsed its thrusters three times to back off a safe distance from the ISS before commencing its deorbit burn. Following splashdown in the Pacific Ocean, a recovery team retrieved more than 1,500 kilograms of samples and human research items sponsored by NASA and the Center for the Advancement of Science in Space (CASIS), the organization that manages research on board the American section of the station. With Dragon out of the way, the next item on the agenda was making sure Bigelow's BEAM was working.

The first attempt to inflate what some hopelessly uninformed media agencies had mistakenly referred to as an 'inflatable tent' was abandoned on May 26, when the module failed to expand as it was supposed to. The inflation team, led by Jeff Williams, spent two hours trying to inflate the BEAM, but succeeded only in increasing the habitat's width. NASA deferred operations for a day to try to figure out what the problem was. The fact that the inflatable had not inflated on the first attempt was not a major surprise, because even Robert Bigelow himself had concerns about how the module would perform. But on May 28, the green light was given and here is how the inflation played out.

The mostly automated process of inflating the structure is executed using a deployment controller located inside the station, with occasional manual input from the crew. The first step in the inflation was to open the Node 3 CBM. This was followed by closing the ascent vent valve, which had ensured the BEAM had been maintained at vacuum since it launched. Next, the restraint straps were released by Williams firing NASA Standard Initiators, which allowed the module to expand a little, whereupon the inflation began. This process, in which air was blown into the bladders between the layers of material, was performed using air from the station rather than from air in the BEAM's tanks, because engineers calculated that the latter option would cause the BEAM to expand too fast, thereby placing excessive loads on the ISS. To release air from the station's tanks, Williams opened a Manual Pressure Equalization Valve in increments over the course of seven hours. Gradually, ever so gradually, the BEAM took shape, until finally it reached its full expansion. Now that it was inflated, the plan was for the BEAM to undergo an 80-hour leak test followed by two years of testing, utilizing myriad sensors.

Since the BEAM is a technology demonstrator, no astronauts will actually live in the module, but crewmembers will enter the BEAM periodically to take temperature readings and clean up dust bunnies. Then, after two years, BEAM will be de-orbited to burn up in the atmosphere. The future? Well, Bigelow would like one of his BA-330s to be attached to the ISS by 2020. That module, called XBASE (eXpandable Bigelow Advanced Station Enhancement), would be a welcome addition to the station which, by 2020, will be staffed by seven crewmembers, courtesy of commercial crew vehicles that should be operational by then.

EXPERIMENT PACKAGE

With BEAM fully inflated, the crew looked forward to the station reboost. That occurred on June 8 thanks to the Progress MS-02, which burned its engines for nearly four minutes using 70 kilograms of propellant. Nearly a week later, it was time for another vehicle to depart. ATK's Cygnus OA-6, which had been attached to the station since March 26, finally departed its berth on June 14. As with the Dragon's departure, the Cygnus, which was berthed on Unity's nadir port, was manipulated using the Canadarm under the guidance of operators Kopra and Peake. Then it was back to the ISS task jar. For Peake, this meant checking in on some of the studies he was tasked with overseeing during his time on orbit. To give you an idea of the scope of these experiments, what follows is a 'Top Ten' of the investigations Peake was involved in during his mission. We'll begin with life sciences.

1. ESA's Airway Monitoring Experiment
The justification of this experiment (Figure 7.8) is based on the respiratory problems that will be faced by those heading beyond Earth orbit. That's because when astronauts finally start exploring Mars (possibly by SpaceX in the late 2020s or by NASA in the 2060s), dust will be a problem. Toxic dust. And since dust may cause airway inflammation, the dust needs to be monitored, hence this study. Very simply, every time you exhale, you breathe out waste-products such as carbon dioxide and nitric oxide, the latter of which can be used as a signaling molecule; in patients, doctors can measure the amount of nitric oxide exhaled to diagnose lung inflammation, and the same principle can be applied to astronauts to chart their lung health.

2. Early Detection of Osteoporosis in Space and Muscle Biopsy Experiments
Extended exposure to microgravity has a devastating effect on bone mineral density. Astronauts lose up to two percent of their bone mass per month and the skeletal system never adapts to a reduced gravity environment, making this system perhaps the biggest mission-killer for those planning to venture beyond Earth orbit for any length of time. In this particular experiment, Magnetic Resonance Imaging (MRI) scans were conducted of Peake's bones during the mission, to determine the structural integrity and strength of his bones before, during and following his time on orbit. In addition to the MRI scans, the British astronaut's muscles were tested for signs of muscle atrophy periodically by inserting a wide-bore needle to extract a small hunk of muscle tissue. Ouch!

3. Skin-B Experiment
While there have been dozens *and dozens* of bone and muscle experiments flown on ISS over the years, very little research has investigated the effect of microgravity on skin. It is

Figure 7.8. Conducting the Airway Monitoring experiment in the Quest Airlock. Credit: ESA/ NASA

known from the few studies that *have* examined astronauts' skin that their skin ages faster in space. Not only that, but skin in space becomes more fragile and takes longer to heal. To help ESA researchers with their Skin-B experiment, Peake conducted a number of tests which we'll go through here to give you an idea of the scope of these experiments and the time demands that just one science payload can impose on an astronaut's workload. We'll begin with the hydration test that Peake conducted using a Corneometer, a device that measures skin moisture. He also monitored his water loss using a Tewameter, a device that measures the evaporation of water on the skin, and evaluated the surface of his skin using a device called VisioScan. It sounds simple, doesn't it? Perform the test and move on to the next one, right? But each test – and remember this is just one of *dozens* that Peake was involved in – necessitates following a precise sequence of steps and instructions. In the case of the VisioScan, Peake was required to place a measuring sensor/CCD-camera on his skin, without pressure, and to ensure the head of the camera rested evenly on the surface of his skin so an image could be captured. Not easy in microgravity. On his return to Earth, Peake was subjected to several additional dermal tests, including biopsies using multiphoton tomography, and assessment of his skin elasticity using a cutometer.

4. Energy Experiment

Along with bone loss, muscle loss, vision impairments, and the catalogue of other bad things that happen to astronauts as they spend time in space, the problem of energy loss can be added. And with energy loss comes loss of body mass, which means all physiological functions are affected to some degree. Why astronauts lose body mass is a mystery, hence the Energy Experiment that measures changes in energy balance during extended duration spaceflight. Peake, for his part, measured changes in his energy expenditure during his mission, to help scientists better understand human metabolism and thereby concoct the ideal diet for those spending months in microgravity.

5. Brain DTI Experiment

The Brain DTI experiment deals with *neuroplasticity*, which is a term that describes the ability of the brain to form new neural connections, thereby reorganizing itself. This is a particularly neat trick when faced with having to adapt to gravity transitions and the effect these transitions have on the body; space motion sickness and cardiovascular impact to name just two. To investigate this ability further, ESA scientists decided to try to track which regions and tracts in the brain were involved in the brain's reorganization. How? They used a non-invasive technique called Diffusion Tensor Imaging (DTI) that measures diffusion characteristics of water molecules in brain tissue, allowing scientists to map the neuro-anatomy of the brain. This technique, when used in tandem with MRI, reveals changes in an astronaut's brain during a long duration mission. While the actual scans were performed pre- and post-flight, Peake was still required to keep a record of headaches and brain-related symptoms during his stay on the station.

6. Circadian Rhythms Experiment

Astronauts don't sleep too well up there. One reason may be the noise, another may be the confined sleeping quarters and another may simply be the fact that sleeping in microgravity is far from natural. But some researchers suspect circadian misalignment may be the culprit preventing astronauts from enjoying restful sleep. Here on Earth, unless you happen to work on a nuclear submarine, circadian rhythms are synchronized by light exposure. Take away that light, and circadian alignment goes out of whack, leading to shortened sleep duration together with cognitive impairment. Not a good mission scenario in other words. Scientists have known about the circadian misalignment problem for years and have conducted studies on the Shuttle, on *Mir*, and more recently on the ISS – albeit on only a handful of astronauts. The studies have found that astronauts just don't adapt to their work-sleep cycles on orbit. Making matters worse is the requirement to maintain wakefulness during adverse circadian phases, due to mission demands such as docking a cargo vehicle. Such a scenario occasionally requires ISS crewmembers to 'slam-shift', a procedure whereby astronauts adjust their sleep-wake cycle to the right or left of their usual bedtime so their wakefulness aligns with mission events. It is not an ideal scenario, and one that can induce circadian misalignment.

Now you might think that's why we have checklists and automated systems, right? Well, checklists are all well and good if you have a crew that is switched on and awake. But if that crew is tired, all the checklists in the world won't make the slightest bit of difference. Take Asiana Airlines Flight 214 in 2013. It's one of the accidents I teach on my *Human Factors in Aviation* course. On an approach to San Francisco Airport, the Asiana 214 Boeing 777's landing gear hit the seawall as the crew attempted an aborted landing. Three people died and 187 were injured, 49 seriously. Why? One reason was because the time of the accident aligned with the circadian low point for the pilots on board the aircraft, a situation that led to compromised decision-making and loss of situational awareness.

NASA is aware of the problem of circadian misalignment, which is why the primary operator is always backed up by a Mission Specialist during docking operations and space-walks, whose job it is to identify and prevent errors. The problem is – and this is based on recent research – that those back-ups are also likely sleep-deprived[2] and suffering from circadian misalignment. Now while this is a problem, astronauts working in LEO still have an extra level of redundancy in the form of mission controllers, who can double-check decisions made by the crew on orbit. That won't be the case en route to Mars due to communication delays, hence the need for this study. So, what did Peake do in support of this experiment? Well, he wore Thermolab, a sensor that monitors temperature and melatonin, a sleep-related hormone. This information, together with the information Peake recorded in his sleep log, will be used to estimate his circadian phase using a mathematical model.

7. Immuno-2 Experiment
Microgravity plays havoc with the immune system. The long duration exposure to radiation, the constant isolation, the circadian misalignment, and the resetting of every physiological system, results in alterations in immune function that ultimately lead to unfavorable health outcomes. The problem with the immune system is that it is *extremely* complex, interconnected as it is to other systems of the body. Stresses such as those imposed on the immune system during spaceflight can suppress all sorts of immune system responses, resulting in compromised wound healing, increased susceptibility to illness, and a long list of other effects that disrupt healthy physiological balance. And remember, this is just a result of spending time in LEO. What happens to the immune system when astronauts finally venture beyond Earth orbit (BEO)? Best get a grip on this immune system compromise, but what exactly happens to cause all these problems? Here's a primer. When the body is subjected to stress, it releases stress hormones such as cortisol, which plays a key role in the activation of the hypothalamic-pituitary-adrenal (HPA) axis. The HPA plays two important roles, one of which is controlling reactions to stress, and the other is the regulation of physiological functions such as digestion. Scientists can get an idea of what is happening in the HPA by measuring cytokines (proteins that signal activation of an immune response) and endocanniboids (substances that regulate physiological functions) in addition to collecting hair, blood and saliva samples, which is what Peake did in support of the study.

8. Spacecraft Fire Experiment - SAFFIRE
Understanding how fire spreads in the confines of a spacecraft or a space station is a critical life support issue, which is why NASA has conducted flame studies on board the Shuttle and the ISS. But until 2016, these studies had been limited in size and scope due to the inherent risks to the crew. So, to avoid endangering the crew, scientists devised an experiment that could be conducted away from the station. That experiment, known as the Spacecraft Fire Experiment, or SAFFIRE, hitched a ride on board the Cygnus that arrived

---

[2] Now there may be some who think the problem can be solved through the use of sleep and wake-promoting medications and you would be right. Sort of. At the time of writing, sleep medication use is almost 20 times higher for astronauts than is consumed by those living on Earth. In most cases, that medication is taken when an astronaut is suffering circadian misalignment, suggesting that other countermeasures (light, or chronobiotics such as melatonin) are simply not working.

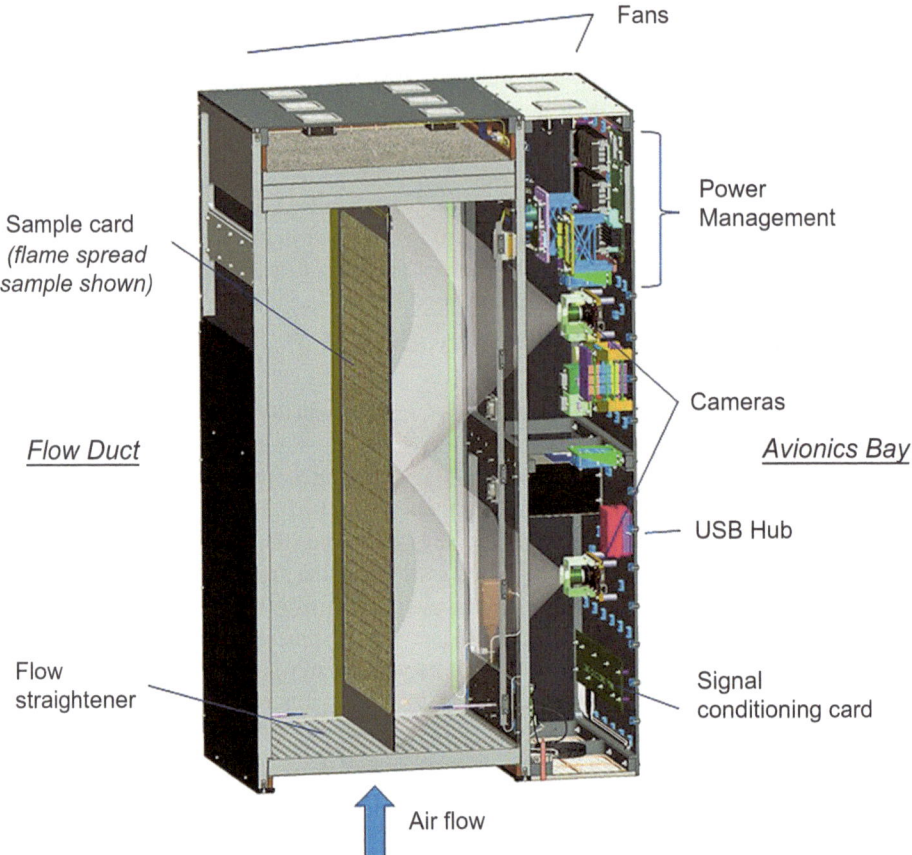

Fans

Power
Management

Sample card
*(flame spread
sample shown)*

Cameras

Flow Duct

Avionics Bay

USB Hub

Flow
straightener

Signal
conditioning card

Air flow

Figure 7.9. The SAFFIRE payload. Credit: NASA

at the station in March 2016. After docking, the astronauts removed all the cargo, except the SAFFIRE payload, which comprised an enclosure as depicted in Figure 7.9.

The enclosure housed a flow duct in which a section of cotton-fiberglass fabric measuring 0.4 meters in width and one-meter in length was set to burn. Once the Cygnus was a good distance from the ISS, the piece of fabric was ignited by a hot wire, and a suite of sensors and video cameras kicked into action to monitor the test conditions; temperature readings were taken, along with measurements of oxygen and carbon dioxide concentrations, and a pressure-time history was recorded thanks to a pressure transducer. While this experiment investigated flame spread and flammability, NASA's plans include additional SAFFIRE missions to investigate smoke propagation, along with the detection and suppression of fire.

9. Gardening in Space. Tim Peake's Rocket Science experiment

"Growing food to supplement and minimize the food that must be carried to space will be increasingly important on long-duration missions. We also are learning about the psychological benefits of growing plants in space – something that will become more important as crews travel farther from Earth."

Shane Topham, Space Dynamics Laboratory engineer,
Utah State University

The life support system on the ISS is classed as a partly-closed system, operating at about 60 to 70 percent closure on a good day. That means lots and lots of consumables must be ferried up on a regular basis. Oxygen, water, spare parts ... and food. It all adds up to an awful lot of weight, and extra weight isn't something you want to be lugging around on a trip to the Red Planet. So, over the years, scientists have been trying to grow food on the orbiting outpost to develop procedures that will optimize food production for BEO missions. It turns out that growing food in space is anything but easy. Fertilizers in space have a slower and more even release rate, roots don't grow in the same way they do on Earth, and artificial light plays havoc with the growth rate of some plants. But with the cost of ferrying just one apple to the ISS coming in at around $2,500 (yes, that's the cost of just ONE apple), growing food in situ is the only option on the table for those hoping to venture beyond Earth orbit. While there is a history of testing plant growth in space, until very recently the goals of that testing were academic. In fact, the first space-grown food that was actually eaten by astronauts on orbit was red lettuce. That momentous first took place in a much-hyped media event in August 2015. After lettuce, space farmers hope that radishes, snap peas, and tomatoes may follow – all designed to take up as little room as possible. But room isn't the only characteristic space farmers are interested in; growth cycles and processing time are important factors too. Leafy greens are ideal because they can be eaten as soon as they are picked from the soil, whereas wheat and rice take longer to grow and require more processing time. As of 2017, we are a long, *long* way from a bioregenerative agricultural capability on orbit. Lettuce might be a nice supplement to the mostly irradiated and thermostabilized food that is served up there, but it is not a long-term food solution, hence ESA's experiment. The gardening project Peake conducted – dubbed Rocket Science – was a comparison of rocket seeds (two million *Eruca sativa* rocket seeds to be precise), some of which were distributed to children around the UK, and some of which accompanied the astronaut to the space station. The objective of the investigation was to see how space – especially radiation – affected the germination and growth of the plants in space.

"Radiation is the biggest risk for dormant seeds stored for a long time in space because cosmic rays are so energetic. When they strike the International Space Station, there is a shower effect when they fragment and become even more energetic. They can penetrate the Station and interact with any biological material onboard. They either pass through the seed or deposit energy there. They can fragment further once inside the seed, too, which leads to greater damage."

Jason Hatton, ESA Life Scientist

The outcome of the Rocket Science experiment was that the space seeds survived the rigors of their journey, since the seeds were determined to still be viable. On that mark, the experiment was deemed to have been a success, but the success was twofold, given the impact the venture had on schoolkids around the UK.

"I hope that this project and my Principia mission have inspired you to study STEM subjects throughout your time at school or college. There are many exciting and rewarding careers out there waiting for you. If you work hard and aim high, there is no reason why you cannot achieve your dreams. I want to say a huge thank you to every single young person who took part in this project. I had great fun watching your photos and messages pour in via Twitter while I was working on board the International Space Station. It was brilliant to see you all enjoying being part of such a fascinating science experiment – and a spot of gardening too!"

Tim Peake

10. Bridget: The Remote Driving Experiment
On April 29, 2016, Britain's eighth astronaut performed yet another high profile challenging experiment, when he remotely controlled a robot rover on Earth (Figure 7.10). The robot in question was named Bridget, and it was located in an oversized sandpit in Stevenage, near London, also known as the 'Mars Yard' (owned by the Airbus Defence and Space Company). The exercise, part of ESA's Meteron (Multi-Purpose End-To-End Robotic Operation Network) program, was designed to assess how well astronauts can control remote systems on other planets. That's because when astronauts finally land on other planets, they will build habitats using remotely-operated robots. Not only that, but it will be remotely-operated robots that will be tasked with almost all surface exploration. That's because humans will be hunkered down in underground bunkers covered by several meters of regolith or Martian soil to protect them from lethal radiation. Forget all those Hollywood scenes of astronauts bouncing across the surface of Mars or all those CGI videos of astronauts exploring the surface of the Moon. Robots will be doing that. Not humans. But let's get back to what Peake was up to with Bridget.

"The Meteron programme is trying to address the end-to-end operations of various robotic platforms by humans, all through robotic systems that can actually take decisions by themselves. To test various scenarios and to validate the related technologies, robots and rovers on Earth will be controlled from the International Space Station with haptic feedback and video footage."

Elie Allouis, Co principal Investigator, Meteron project,
Airbus Defence and Space team

Peake's task was to remotely drive Bridget in a dark environment; that environment being the aforementioned Mars Yard in Stevenage. The reason for the challenging light conditions was that daylight doesn't present a problem for the latest generation of rovers, because they are fitted with autonomous navigation systems that allow them to plot a route and drive it just fine. But when the lights go out, battery life becomes an issue, and having a rover controlled by an astronaut means the rover can overcome unexpected obstacles more effectively. As long as said astronaut can control the rover. Hence the Bridget test. Using Bridget's sensors, Peake had to identify painted targets in the Mars Yard that represented a cave on the

Figure 7.10.   Driving rovers. Credit: ESA

Red Planet, a task that sounds simple enough until you consider the fact that there was a delay before his instructions took effect because of the several seconds it took to relay the signal from the ISS. Thanks to some nifty maneuvering, Peake was able to log some quality exploring time and navigate his way out of the cave before handing back control of Bridget to ESA's European Space Operations Center in Darmstadt, Germany.

BACK ON EARTH AFTER 186 DAYS IN SPACE

Peake's return to Earth began in the same Soyuz TMA-19M spacecraft that took him to orbit on December 15, 2015. By the time he squeezed into the Soyuz in the early hours of June 17, 2016, he had orbited the Earth 2,720 times and had traveled 114,240,000 kilometers. The Soyuz hatch closed at 3:34 am (UK time) on June 17, thereby marking the official end of Expedition 47 and the Principia mission. Ensconced within the claustrophobic confines of the Soyuz were Peake, Kopra and Malenchenko, crammed together side-by-side in special custom-fitted shock-absorbing seats. At 6:52 am, the Soyuz undocked from the ISS and performed a short, 15-second burn to edge away from the orbiting outpost. Landing was scheduled for 10:15 am. After performing the four-minute 45-second deorbit burn, the Soyuz, still zipping along at 28,000 km/h, began its plunge through the atmosphere. Thirty minutes before landing, at an altitude of 140 kilometers, explosive bolts fired and split the Soyuz into its three elements. The Descent Module – the middle part containing

Figure 7.11. Tim Peake phones home after landing. Credit: ESA

the crew – flipped over to point its heat shield in the direction of re-entry, while the Service Module and the Orbital Module burned up in the atmosphere. As the atmosphere began to bite, the outside temperature edged upwards to 1,600°C and the deceleration pushed the crew back into their seats, as the 'Gs' topped out at 5G.

> "It was incredible. The best ride I've been on ever. Truly amazing. A life-changing experience. The smell on Earth is really strong. Looking forward to seeing the family. I'm going to miss the view, definitely. I'd like some cool rain right now; it's very hot in the suit. It's very hot in the capsule."
>
> *Peake, on returning to Earth, before going on to say*
> *he fancied a cold beer.*

At 10:00 am, radio communication was lost as the Soyuz was enveloped in a shroud of plasma. When communication was re-established at 10:05 am, the first transmission was from Malenchenko, who informed Mission Control, "We feel good." This was followed two minutes later by confirmation that the parachutes had opened at an altitude of 10,800 meters. At 10:10 am, the crew was instructed to strap themselves in for the landing. Touchdown, just eight kilometers from the planned site, occurred at 10:15 am. Twenty minutes later, the three crewmates had exited the capsule and were wearing smiles while lying on the familiar recovery couches, as medics scurried around testing oxygen saturation and blood pressure. A few minutes later, Peake gave the thumbs up and then did what all explorers do on returning from an expedition – he phoned home. Snippets of what was said appeared on Twitter, with Peake's parents, speaking from ESA headquarters in Cologne, saying they were very calm, although the previous few hours had been nerve-wracking.

After spending some time soaking up the experience of being back on Earth, Peake and his crewmates were taken by helicopter to Karagandy airport where, according to time-honored Kazakh cosmonaut tradition, they were offered bread-and-salt and a Kazakh hat. From there, the crews went their separate ways, not even waiting for the post-mission press conference. For Peake, his next destination was ESA's Astronaut Centre in Cologne, via a stopover in Bodø, Norway, courtesy of a NASA Gulfstream jet. He spent his first night back on Earth at the Envihab facility of the DLR German Aerospace Centre in Cologne. There was no word as to whether he listened to Peter Schilling's 1983 hit 'Major Tom'.

Notable British-themed mission events
Second British national, after Helen Sharman, to enter space
First "official" British astronaut, sponsored by the UK Government
Observed the 90th birthday of Queen Elizabeth II
Ran the London Marathon on the station's treadmill
Celebrated St. George's Day

Notable numbers
Delighted a worldwide audience with #SpaceRocks – his selection of 75 favorite music tracks, whose opening lyrics appeared regularly on his Twitter account
Held a science lesson for 300,000 schoolchildren.

1,672,203. The number of social media followers Peake had on Twitter, Facebook, Flickr and Instagram.

117 songs. The number of tunes Peake had on ESA's Spotify account. One particularly apt listening choice: *Way Down* by Tom Odell.

Notable dates

December 24, 2015. Sorry, wrong number. "I'd like to apologize to the lady I just called by mistake saying 'Hello, is this planet Earth?' – not a prank call ... just a wrong number!" Pensioner Betty Barker, 79, receives an unexpected call from the ISS after Peake dials the wrong number and inquires 'Is this Planet Earth?'

January 15, 2016. Britain's first spacewalk.

January 16, 2016. First flower grown in space. The first flower blooms in space under Peake's green-fingered care.

February 24, 2016. Brit Awards. Peake reduces Adele to tears after presenting her with a Global Success award.

June 10, 2016. Queen's Honours. Peake is awarded the Most Distinguished Order of St. Michael and St. George in the Queen's Birthday Honours for extraordinary service beyond our planet.

Mission records

Malenchenko logged 828 days in space, making him second on the all-time list behind Russian cosmonaut Gennady Padalka, with 878 days.

Kopra logged 244 days in space on two flights

Peake spent 186 days in space on his first flight.

8

So how did he do?

Figure 8.0. Credit: ESA

THE TIM PEAKE EFFECT

"Truly elated, the smells of Earth are just so strong, just so good to be back on Earth. I'll look forward to seeing the family. It is going to be quite tricky for me to adapt. It's probably going to take me two or three days before I feel well. It will take me

© Springer International Publishing AG 2017

E. Seedhouse, *Tim Peake and Britain's Road To Space*, Springer Praxis Books,

DOI 10.1007/978-3-319-57907-8_8

several months before my body fully recovers in terms of bone density. And it will be interesting to see any lasting changes to eyesight etc. But generally speaking, in two or three days I should be fairly comfortable back on Earth."

Major Peake's comments in his last news conference
before the return to Earth.

"He's done an amazing job. He's exceeded all expectations, certainly in terms of the impact that he's had back in the UK."

Dr. David Parker. The British Space Agency's
director of human spaceflight

THE MEDICAL ISSUES

In a matter of moments following his return to Earth, Tim Peake transformed from an orbiting Superman to a convalescent in need of months of rehabilitation. It's an adjustment every long duration astronaut has to make; after somersaulting with the ease of an Olympic gymnast and wielding payloads with their fingertips, astronauts returning to Earth's gravity after a long stint on orbit are faced with numerous physiological challenges that take months *and months* to get to grips with. Returning astronauts have lost bone mineral density, suffered visual impairments, and become discombobulated due to neurovestibular changes and adaptations; many have trouble standing upright due to their balance muscles having wasted away, and are prone to infection due to spaceflight-induced immune system compromise (Figure 8.1). The list is endless. In fact, the effects of long duration spaceflight are so severe that those returning from their stay on orbit are banned from driving for three weeks.

Fortunately, a small army of strength/rehabilitation specialists and flight surgeons are on hand to help astronauts through the long haul required to adjust to one G, a process that requires spacefarers to submit to tests. Lots and *lots* of tests. Some of these tests are scientific and form part of the data set of larger studies, while other tests are administered to assess the extent to which the astronauts are meeting recovery standards. Doctors draw blood, conduct magnetic resonance imaging (MRI) scans, submit endless questionnaires, place astronauts on tilt tables, perform muscle biopsies and ... well, you get the idea. But why all the tests? Because when astronauts spend any time in space their bodies deteriorate, and the longer they spend, the greater that deterioration is. Here's a snapshot of some of the effects that Peake's mission wreaked on his body.

Tim Peake nursing 'world's worst hangover' after six months in space

British astronaut experiencing dizziness and vertigo as he readjusts to Earth and begins intensive rehabilitation in Germany.

Tagline from the British newspaper, The Guardian, *June 20, 2016.*

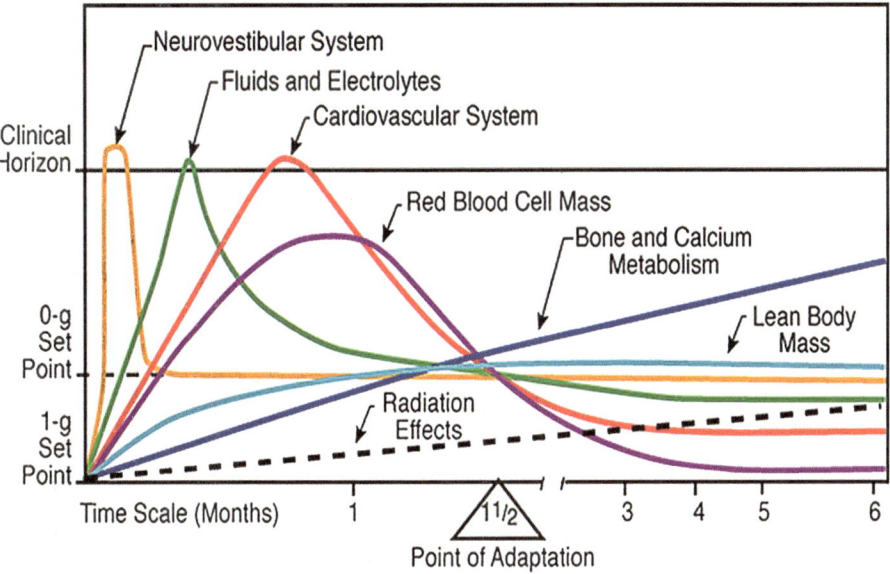

Figure 8.1. Graph showing adaptation of physiological systems. Credit: NASA

Fluids

Let's begin with the effects of fluid shift. You've got lots of fluid in your body; blood, bile, gastric acid, pericardial fluid, pleural fluid, lymph, serous fluid, and urine to name just a few. And when an astronaut enters a microgravity environment, between one-and-a-half and two liters of that fluid shifts towards the head and chest. This causes all sorts of problems, the most benign of which is the feeling of stuffiness and feeling as if you're coming down with the flu. A bigger headache is the increased intracranial pressure caused by all that extra fluid in the skull cavity. This extra fluid exerts pressure on the physiological structures, including the eyeball and optic nerve. So much pressure is exerted that the shape of the eyeball is distorted. As you can imagine, if you begin messing around with the shape of the eyeball or start crushing the optic nerve, vision may be compromised, and that is exactly what happens in some astronauts (Figure 8.2). Another problem with all this fluid rushing to the *cephalothoracic* region (a fancy term space life scientists use to describe the movement of fluid from the extremities to the head and chest) is that the brain registers extra fluid pressure courtesy of pressure sensors (*baroreceptors*) strategically placed in the body. The brain's response? Get rid of the extra fluid. The result is *diuresis*, which basically means the body begins to purge itself of fluid by frequent visits to the washroom. Now you may be thinking that the body adapts, and it does, because the body is an extraordinarily versatile system, but take a look at Figure 8.1. Closely. Notice anything about that second horizontal line? That line is the 'zero-G' adaptation. Great if you're a long duration astronaut. Not so great for when you return to Earth.

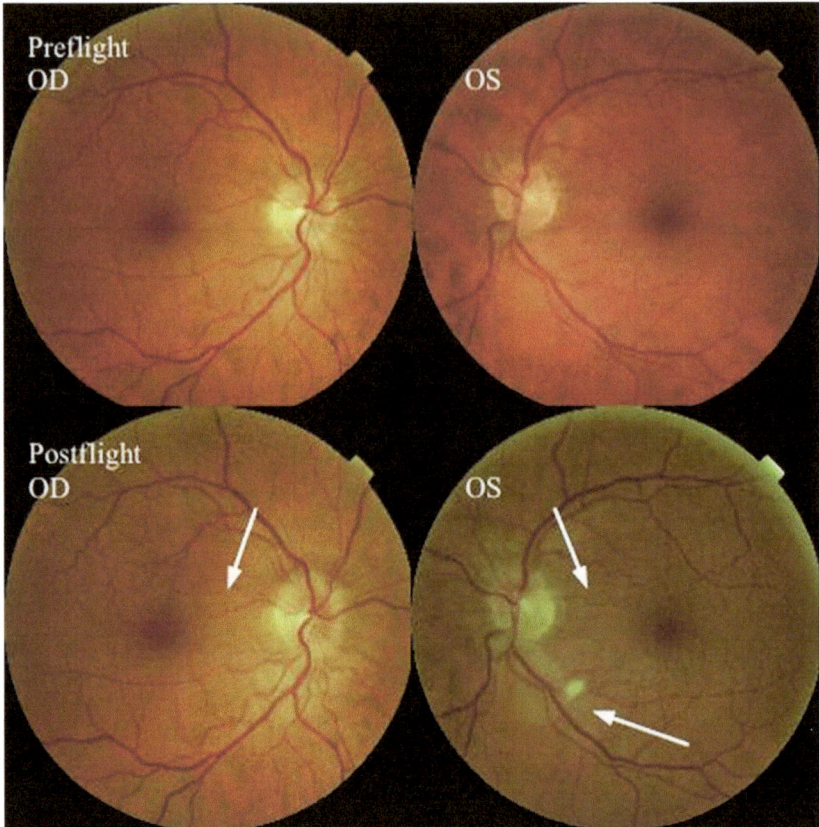

Figure 8.2. VIIP. The effect of weightlessness on eyesight has a long history stretching all the way back to the Mercury program. While vision tests performed during the Gemini V and Gemini VII missions showed very little change in an astronaut's visual acuity, similar investigations conducted during the Apollo program noted an increase in intraocular pressure (IOP). The Apollo findings were later confirmed by Spacelab studies, which revealed an in-flight IOP increase as much as 25 percent higher than pre-flight. The results of comparable studies performed in the Shuttle era echoed earlier findings, but the ophthalmic changes didn't have a profound effect on vision. Then, in 2005, the visual impairment/intracranial pressure (VIIP) problem surfaced, which was the catalyst for researchers performing the retrospective survey of questionnaires that were submitted to 300 Shuttle astronauts. Some crewmembers, all of whom were males between 45 and 55, experienced symptoms so severe that some had difficulty reading checklists. And one astronaut, who suffered scotoma, had to tilt his head 15 degrees to view instruments: his symptoms were still evident more than a year after landing. Something had to be done, hence the vision studies being conducted on ISS today. Credit: NASA

Muscles

So that's the fluid system. Let's move on to the muscles. As soon as astronauts enter zero-G, the body begins to react to its new environment. With no gravity, astronauts don't have to work as hard as they do on Earth, which is another way of saying the muscles don't have to work as hard. Since astronauts spend most of their day floating around, one of the first muscle groups to take a hit are the postural muscles, which begin to atrophy. As do the leg muscles, which are normally used for walking, and the heart muscle, which also begins to wither now that it doesn't have to work as hard. Now, you will have seen videos of astronauts sweating away on treadmills and bike ergometers, and probably assume that all that exercise must surely be an effective countermeasure against all that muscle loss. Well, it is, but two hours of exercise can only *slow* the process of muscle atrophy, not eliminate it, which means astronauts who spend six months up there may lose up to 30 percent of their muscle mass. Thirty percent! And you wonder why the astronauts are lying down on couches when they return from their stint in space.

Bones and radiation

Now to bone loss (Figure 8.3). Here's a real mission-killer for those planning trips to Mars. Take a look at Figure 8.1 again. Notice anything about the blue line that's heading off the page to the upper right of the diagram? That's the skeletal system. It's the one system that doesn't adapt to spaceflight. Astronauts just keep losing more and more and *more* bone. Even by sticking to a rigorous exercise routine on orbit, astronauts still lose between 1.5 and 2 percent of their bone mass per month. Per month! That is ten times the

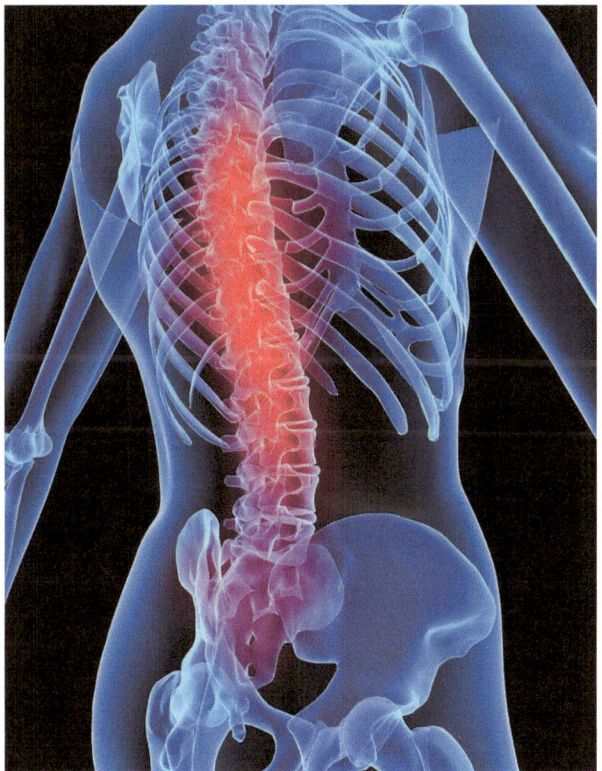

Figure 8.3. Spinal x-ray. Credit: NASA

rate of someone in their 70s. Do the math: after six months on orbit, astronauts may have lost about ten percent of their bone mass. Another reason for those couches! And the greatest loss is in the femur and the pelvis, which increases the risk of injuries such as a hip fracture. Imagine an astronaut who's just spent the best part of six months getting to Mars and they step onto the surface of the Red Planet and their femur snaps! Remember – no rehab teams waiting on the surface of Mars. But we're off topic. Radiation is next (Figure 8.4). During his time on the ISS, Peake was exposed to the radiation equivalent of 1,200 chest X-rays. Nothing to worry about in the short-term, but that amount of radiation has been calculated to increase the risk of cancer by as much as three percent.

So that's a snapshot of some of the physiological challenges Peake had to contend with on his return. The good news is that many of the changes are recovered in a matter of months. The bad news is that, in some cases, bone loss takes many years to recover. In that

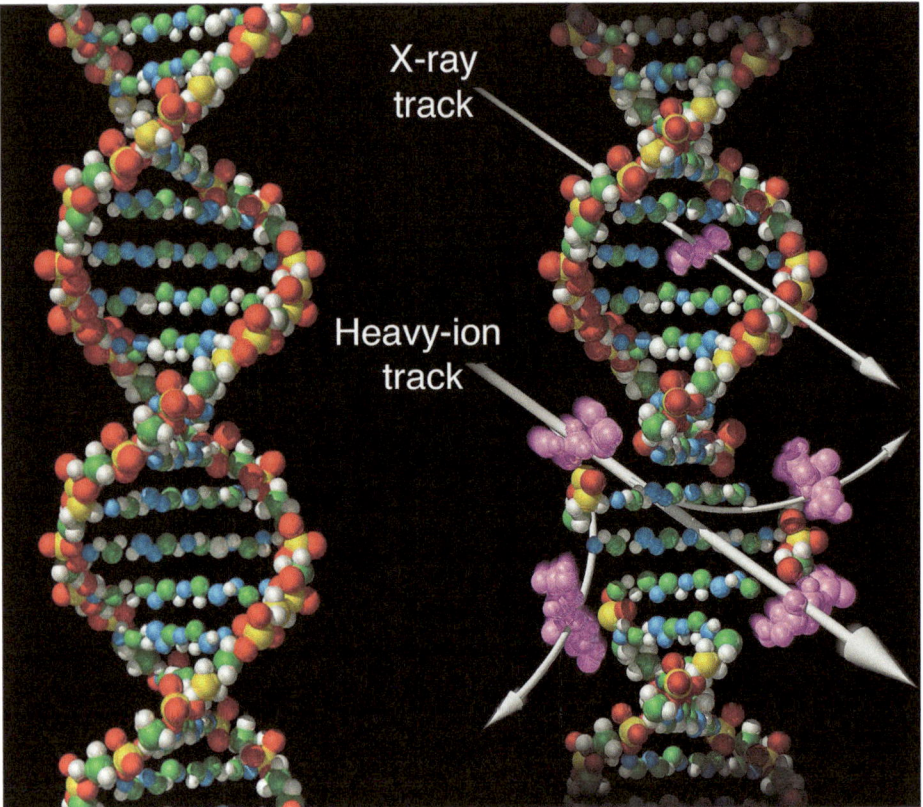

X-ray
track

Heavy-ion
track

Figure 8.4. DNA. Why investigate DNA? Well, when astronauts live on board the ISS, they are exposed to radiation. A lot of radiation. Down here on Earth, you will be exposed to an annual dose of 2 milli-Sieverts (mSv). On board the ISS? 80 – yes, EIGHTY – mSv. And that's just in six months. Not excessive, but that number jumps up dramatically once you travel beyond low Earth orbit, and that's going to be a problem for those venturing to the Moon or Mars because it's very difficult to shield against radiation such as galactic cosmic rays. Credit: NASA

regard, astronauts are guinea-pigs for experiments that may one day help humans finally leave low Earth orbit (LEO) once again, but in the case of Peake, his impact on the UK is probably more important. Or is it?

SUCCESS, OR FOLLY?

While he grabbed the headlines as Britain's first government-funded astronaut, addressed the Brit Awards, and ran the London Marathon in space, what did Tim Peake's mission add up to? Most media coverage following his mission was positive, but there was the odd pundit who argued that the venture was little more than political vanity. Let's examine that argument first. The naysayers who don't see the value of human spaceflight argue that the money could be better spent elsewhere, such as robots and satellites. These wet blankets go on to argue that since Peake was only the eighth Briton in space, depending on how you determine citizenship, and only the 373rd person to visit the ISS, the Tim Peake story is no big deal. The argument against spending money on human spaceflight is supported by the costs incurred by the American government in keeping the Shuttle fleet in orbit, at a figure north of $1 billion per flight (when all development, maintenance and operating costs are considered). Former UK Prime Minister Tony Blair decided against blowing money on the successor to Shuttle, the ISS, whereas David Cameron decided to spend £16 million on the station by way of the European Space Agency (ESA), which was the catalyst for Peake's flight.

For the 'use robots, not humans' faction, Cameron's decision was borne of vanity and prestige. That's because the main purpose of the ISS is to serve as a platform for research to support expeditions to Mars, and in the minds of the 'robot supporters', a trip to Mars is the highest form of political vanity, given that such a trip is pointless and perhaps even suicidal, according to the doom and gloom merchants. To bolster their argument, these prophets of doom point to other prestige projects that came to ignominious ends, such as high-speed trains and *Concorde*. It is a powerful argument, when set against the line of reasoning posited by the space lobbyists, who cite unquantifiable intangibles such as the spirit of exploration and the legacy of discovery and adventure. Then there are the politicians who, in the case of Peake's flight, were responsible for the Principia mission. Not surprising then that a blue-suited Peake found himself in Number 10 together with David Cameron shortly after the ISS mission had been announced. The question is how the UK approaches space. On the robotics and satellites side, it seems to be just a business venture. After all, robots (see sidebar) and satellites are just more industries and there's very little magic in that. When was the last time you were inspired by a robot? But manned spaceflight is a different animal. The problem is Britain has only had the experience of flying one astronaut, so it's difficult to gauge exactly how it plans to exploit this newfound opportunity.

Robots versus humans
The debate about the merits of exploring space with robots or humans has been around as long as there have been astronauts. Wernher von Braun believed in the value of human exploration, whereas James van Allen was an ardent fan of sending

robots off to explore. So, who's right? Well, let's begin by supporting humans and touch on some of the arguments put forth to send astronauts into the unknown. One hyped up argument is that manned space exploration will help us avoid the fate of the dinosaurs, if such a catastrophe were to come to pass. Another argument is that money spent on manned spaceflight is good for the economy (in the U.S., every dollar spent on space translates to $8 in economic benefits). Then there's the old chestnut of international cooperation as a means of reducing costs and avoiding national hostilities. An even older chestnut is the 'exploration for exploration's sake' argument. Anyone with the flimsiest sense of history knows that great civilizations do not pull the plug on exploration.

"Because it was there. Exploration is intrinsic to our nature. It is the contest between man and nature mixed with the primal desire to conquer. It fuels curiosity, inspiration and creativity. The human spirit seeks to discover the unknown, and in the process, explore the physical and psychological potential of human endurance."

Mountaineer George Mallory's response when asked why kept trying to climb Everest

Robots? Well, they play a part in scouting out potential landing sites for the real explorers, and they do that without risking lives, but who repairs the robots? Humans, of course. But what about the cost of sending humans into harm's way, the robot fans wail? Is it worth it? That's a little like asking if Columbus's trips to the New World were worth it. At the time, the answer was clear to some whereas others took a while to realize the benefits. That's why countries such as China and India, each with appalling social welfare issues, still commit large amounts of money to manned spaceflight and spaceflight in general. But still the robot devotees argue that the money would be better spent on Earth, apparently oblivious to the fact that's precisely where it *is* being spent: just think of all the jobs and companies that exist just because of the space program. And while we're on the subject of value, just think of the inspiration generated by the manned spaceflight program, a type of inspiration that is unique to this industry, motivating hundreds of thousands of people to pursue a career in science. In the absence of a manned spaceflight program, what catalysts exist to drive the next generation to follow science, technology and engineering careers?

A positive influence

The potential comedown following Peake's flight was one of the concerns voiced by the media. They worried that the wave of excitement generated by Peake while on station would dissipate, and the UK would revert to a life as normal shortly after his return. Fortunately, that doesn't seem to have happened. In part, that enduring popularity stems from the fact that Peake has been a regular feature in the headlines ever since his selection in 2009. Press campaigns have helped, as has social media – especially Twitter, which hosts 780,000 followers of the British astronaut, more than any other ESA astronaut. In fact, thanks to the intense public interest in all things Peake, the former Army helicopter pilot has become a household name. For that level of awareness, thanks have to go to the UK Space Agency (UKSA), which whipped up plans for the largest and most extensive education program ever devised in support of an ESA astronaut. The UKSA, in tandem with ESA, who shared details of the Principia mission on as many social media platforms as possible (Twitter, Facebook, Instagram), were the catalyst for the innumerable media requests that poured in. In orbit, the high-flying Brit slipped into a fake tuxedo in honor of the Brit Awards. He called down to schools to answer questions from enthralled school children, got chased by Scott Kelly who had dressed up in a gorilla suit, and misplaced a call to his parents. And always there was that broad grin and the thumbs up. The media loved him.

"The response was amazing. We have a responsibility to share these stories with the people who fund the missions. Now, with this fantastic tool of social media and the internet, we don't have to wait until the astronauts get back home."

Jules Grandsire, communication officer at
the European Astronaut Centre

Of course, the success of the mission wasn't entirely down to the mass media. The most famous man from Chichester would not have amassed quite such a following if he did not possess a certain charisma and charm. Not convinced? Just check out the photos of him saying goodbye to his toddler son, Peake's hands pressed to the glass. In fact, try to find any photo of Peake when he's not beaming like the proverbial Cheshire cat. Peake's antics on orbit brought back memories of Chris Hadfield, another famous Commonwealth astronaut, who made social media his own and set the benchmark by which astronauts nowadays share their experiences while whizzing around the Earth at 28,000 kilometers per hour.

"Tim has balanced it really well. He's made it look accessible and interesting to so many people that otherwise might not have seen it. But he couldn't have done it without the last decade of advances in communications technology. That has really democratized the experience."

Chris Hadfield, Canada's most famous astronaut and
one of its most famous public figures,
weighs in on Peake's social media savvy.

INSPIRING THE YOUNG

While the impact on the media was nothing less than stellar, the influence on education was even more dramatic (Figure 8.5), with more than one million kids learning about space during the Principia mission. Children who interfaced with the famous astronaut told their teachers they would remember the day they spoke with Major Peake for the rest of their lives. How many robots have that effect? A million children across 30 educational projects that centered around the Principia mission ensured that the 'Tim Peake Effect' will continue for years, thanks to the mission's giant leap in creating nothing short of a culture shift on STEM projects across the UK. It's a momentum that the UK cannot afford to lose, especially given the country's position as an industry leader in space technology. It's a position that many people in the UK are unaware of because when people mention

Figure 8.5. Tim Peake with Mission-X kids. Credit: ESA

space, they immediately conjure up images of NASA. But thanks to the endeavours of Tim Peake, people realize the power of science as being not only accessible, but also as a catalyst for inspiring kids. And that inspiration couldn't have been better timed. That space industry is growing and creating more jobs, but without the talent and academic know-how to fill those jobs, the UK's leading status is at risk.

In fact, with forecasts predicting a shortage of 40,000 STEM-related jobs a year by the end of the 2010s, the British space industry was already imperiled before the start of the Principia mission. Before Tim Peake rocketed off the pad at Baikonur, survey after survey revealed that many people had turned away from science and mathematics after their GCSEs because these subjects were considered difficult, and many kids couldn't connect these STEM skill-sets with jobs and future careers. What was needed was a way to show kids that STEM can be accessible and inspiring, so in that regard the on orbit platform that was Principia was the most golden of golden opportunities for teachers up and down the country. Livestream broadcast followed livestream broadcast, as Peake delivered the exciting world of science to thousands of children, many of whom were able to ask the astronaut questions about life on the ISS. The outcome of six months of the highest of educational platforms was nothing less than that cultural shift across STEM, a shift that also created a momentum that many vowed had to be continued, especially in light of previous opportunities lost.

A reminder in case you haven't read Chapter 1: the UK could have done so much more in the spaceflight arena and done it so much earlier, had it not been for political viscosity and some bad luck. Way, way back in the 1950s, more than half a century before Peake launched a new era for British manned spaceflight, the UK built a hush-hush site on the Isle of Wight to develop and test rockets for the purpose of launching satellites. One of these rockets, a Black Arrow, carried Prospero into orbit in 1971. It was the first time a British satellite had been launched by a British rocket. Experts believe that variants of the Black Arrow could have been developed to carry humans, but the option of pursuing the goal of becoming a major spacefaring nation was ditched when Britain's rocket program was cancelled. Instead, Britain became a footnote in the annals of rocket history, as the only nation to have developed a launch capability and then abandoned it. That was the political viscosity side of the coin. The bad luck? That happened in 1980s, when Britain had a chance to send its own astronaut into orbit thanks to Ronald Reagan's Strategic Defense Initiative. That would have been the catalyst for Britain training a corps of astronauts to launch military satellites. Then *Challenger* exploded and that was the end of that opportunity.

But now, in 2017, thanks to Tim Peake and David Willetts[1], the UK is back in the space race. There may be some who argue that Britain was in the manned space race already, thanks to Virgin Galactic and its efforts to launch spaceflight participants on its SpaceShipTwo, but the noise about that venture died following the crash of SpaceShipTwo in October 2014. Virgin Galactic will hopefully rise from the ashes, get their act together and fly fare-paying passengers, but it won't be for many years yet and certainly not before the end of this decade. And while many admire the tenacity of Virgin Galactic in developing space tourism, seeing a Virgin Galactic-branded spaceship doesn't quite generate the same amount of pride as seeing a Union Jack on the arm of a British astronaut.

"ESA believes that human spaceflight delivers excellent science and innovation, is a dramatic demonstration of international partnership in difficult times and, through its astronauts, presents striking role models for the young generation. For example, the education activities around Tim Peake's mission have reached one million young people, perhaps the largest science engagement project ever accomplished in the UK."

David Parker, ESA Director of Human Spaceflight
and Robotic Exploration

Educational impact

But we can't put a number on national pride and people want numbers, so let's backtrack a little and revisit the educational impact of Peake's visit to Earth's orbiting outpost. We've already mentioned those million students, but hidden in that statistic are some other staggering numbers. Here are a few of them. The thirty projects that were developed for the Principia mission comprised the largest public engagement task ever undertaken by an ESA astronaut. Let's take a look at some of these projects, beginning with the Rocket Science Experiment (Figure 8.6). The details of this experiment are provided in Appendix VI,

---

[1] David Willetts was the UK Science Minister from 2010 to 2014, and is generally regarded as the person responsible for overturning Britain's myopic space policy and thereby securing Major Peake's flight.

Figure 8.6.   Tim Peake in the Cupola, with 2 kilograms of rocket seeds. Credit: ESA

so what follows is a snapshot. In this project, more than 8,000 packets of seeds from rocket plants were send to thousands of schools, ultimately reaching around 600,000 children. Half the packets contained seeds that had flown for six months in space and the other half were seeds that had remained on the ground. Students cultivated the plants to observe any differences in growth. The outcome was a research platform that could be used by teachers to explain core principles of research such as randomization and blind trials. In addition to the myriad educational opportunities, there were plenty of live link-ups to the ISS, during which Peake spoke directly with schoolchildren. The Cosmic Classroom was one such event, which was followed by 400,000 watching via a webcast. Destination Space was another event that was staged at science centers up and down the country, attended by 350,000 interested in learning about Peake's mission.

So those are some numbers, but of course those numbers don't amount to a tin of beans if they don't translate into meaningful changes in the way children approach the STEM subjects. It's a question that has inspired unique research at the University of York, where a team is examining if there will be a 'Tim Peake Effect' in the same way there was an 'Apollo Effect' back in the 1960s. The question posited by the Science Education researchers at the university is 'Does space flight inspire school students to take STEM subjects?' The three-year project, which is funded to the tune of £348,000 by the UKSA and the Economic and Social Research Council (ESRC), will gather input from students and teachers from a sample of 30 primary and 30 secondary schools, using an instrument designed to assess attitudes to STEM subjects.

"There is anecdotal evidence to suggest that space and space travel increase the interest of young people in science, technology, engineering and maths (STEM) subjects. We have a golden opportunity to gauge this hypothesis as we prepare to send a British astronaut into space at the end of next year."

Principal Investigator Professor Judith Bennett,
Department of Education, University of York,
commenting before Tim Peake's mission

BRITAIN'S FUTURE IN SPACE

While Peake was an exceptional ambassador for STEM education and for firing up enthusiasm in schoolchildren for science, the direct impact of his involvement won't be known for a few years. But it is clear the 'Tim Peake Effect' should not share the same fate as the Black Arrow. Equally, it will be a while before the outcome of Peake's involvement in the science he conducted will be known. This isn't surprising, given that he was involved in more than 200 experiments, and it is rare for an astronaut to have a pivotal role in any experiment during their time on board the station. But beyond serving as a STEM ambassador, Peake also played a promotional role in highlighting Britain's £10 billion space industry which, until the man from Chichester flew to space, operated largely in the shadows away from public attention. One political figure who took notice of this fact was none other than David Willetts, the former science minister – now a Lord – who was the catalyst for Peake's mission. Here's what Lord Willetts had to say:

"Tim has been a brilliant ambassador. His flight has achieved more than we dared hope. It's hard to measure, but I believe Tim has given Britain some real confidence that we are a serious player in space. People didn't recognize that we design satellites and instruments for Mars and that a Mars rover is being built in Britain."

Such an endorsement should bode well, but nothing is certain in politics, especially when it comes to space. So, one key question is whether there will be more flights for British astronauts. That will depend on the strength of the 'Tim Peake Effect' and if there is money allied with political will. According to the man himself, there *will* be a second flight. On January 26, 2017, Major Peake announced he would be making a second flight to space, although the timing had yet to be decided. While a second flight by a British astronaut, even if it happens to be the same astronaut, would be welcomed by the STEM crowd, there are some who seek justification for the tens of millions of pounds such a venture will cost. It's a reasonable question, so what is the point of manned spaceflight nowadays? Well, it depends. Way back when manned spaceflight was just a theory, those who pursued the idea of sending astronauts into space were pioneers who experimented with crude rockets. These pioneers had names like Konstantin Tsiolkovsky, also known as the Father of Cosmonautics, and Robert Goddard, Hermann Oberth, and of course the Godfather of manned spaceflight, Wernher von Braun. These legends of the space industry forged a path that ultimately led to the start of the Space Age proper, with the launch of Sputnik and the myriad firsts that followed.

The pioneering phase eventually gave way to the geopolitical phase, a chapter in time that aligned with the Cold War that was all about the Soviet Union and the United States trying to prove their superiority over each other. To begin with, the Soviets led the U.S. by stringing together all sorts of manned spaceflight firsts. But ultimately, they were upstaged by the master stroke that was Apollo. In the propaganda that was used to justify the incredible costs of beating the Soviet Union to the Moon, the American public was subjected to fear tactics, and were told the Soviets would launch nukes from the Moon. Fast forward 45 years and the fear mongering has been diluted, but geopolitics is alive and well, with countries cleverly using manned spaceflight to advance foreign policy – just take a look at the ISS as a prime example (for Britain, the main geopolitical significance of Peake's flight was the nation finally losing its status as the only G8 country not to have flown an astronaut on the ISS). The question is whether geopolitics is worth the $150 billion spent on the ISS (a figure that equates to a cost of $7.5 million per crewmember per day, as mentioned). The answer depends on who is forking out the money. For the commercial sector, the reason for manned spaceflight is mostly money, unless you happen to be Elon Musk of SpaceX, in which case pursing human spaceflight is all about boot-prints on Mars. For governments, the reasons to stay in the manned spaceflight game are to advance science, generate economic return, and to advance foreign policy.

Some argue that advancing science for the sake of science is a thing of the past and such ventures should show an impact or have some sort of tangible benefit. For the Brits, the tangible benefit of their Principia mission was primarily educational outreach and inspiring the next generation, and on those counts they were supremely successful. There is no doubt that trying to recruit scientists and engineers to support the economy is a challenge for any country, and there is also no doubt that manned spaceflight is perhaps the most potent recruitment tool in this regard. And the Brits played that card beautifully. Not only that, but Principia firmly placed the spotlight on the British space sector, an industry that employs over 100,000 people and generates north of £10 billion a year. All of a sudden, the British public who, until the Principia mission, assumed space was just manned spaceflight, were being informed by BBC sound-bites that their country uses space for everything from tracking crop yields to tackling illegal fishing. Those sound-bites, together with the Twitter feed from the ISS, helped the British public better understand the benefits of manned spaceflight; benefits echoed in Britain's National Space Policy. The aforementioned 15-page document, which was published by the UKSA on December 13, 2015, just days before Peake was launched to the ISS, opens with the words 'Space matters'. It goes on to describe the roles and responsibilities of the various government entities linked to spaceflight, before taking a more exploration-based tone that suggests the UKSA plans to send more astronauts to the ISS and even land one on the Moon someday:

"The path towards the achievement of the vision begins with space-analogue and orbital platforms, expands human presence into the solar system, and ultimately leads to human missions on the surface of Mars or other solar system planets, moons or asteroids. The first steps focus on utilizing the ISS, expanding the synergies between human and robotic missions, and pursuing discovery-driven missions in the lunar vicinity that help to develop capabilities and techniques needed to go further."

To meet those goals money will be required, but Britain has already pledged £42.9 million to the ISS, which is a start. It may not be sufficient to guarantee a second flight of a British astronaut, which is perhaps why the document is light on the details of future British manned missions, although mention is made of how the Principia mission served as a major source of inspiration for students interested in STEM. The policy does a good job at reaffirming that space does matter to the government, but whether that affirmation results in more British astronauts remains to be seen. You see, the short-term scientific returns of launching astronauts to orbit are pretty slim, which is why the reasons for sending astronauts to space nowadays are more about national and international prestige. Science? Despite all the great work Peake did promoting STEM, the British government's budget lines for science are positively anorexic, since only 0.48 percent of the country's GDP is spent on science (the lowest in the G8). So, sending another British astronaut up to the ISS using cash from the UK science budget is a non-starter. And, since the UK decided to exit the EU, that state of affairs is unlikely to improve.

So, was it worth all the time and money and effort? Look, space, like most things worth pursuing, is a long-term game that requires long-term commitment. If the UK wants to commit to manned spaceflight for the long haul, it has to ensure long-term funding that matches the budgets of other countries that have invested in the manned spaceflight business. If that doesn't happen, then Tim Peake's flight will be relegated to a one-off prestige project. And that would be a real shame.

Appendix I

Astronaut Selection Criteria

How can I apply?

You can apply online via the ESA web portal (www.esa.int/astronautselection - whenever the next selection is announced).

Registration begins with Pre-registration: you will need to provide identity information and a JAR-FCL 3, Class 2 medical examination certificate or equivalent, from an Aviation Medical Examiner who has been certified by his/her national Aviation Medical Authority; or alternatively the ESA Medical Statement, approved by a physician.

What are the steps in the selection process?

1. initial selection according to basic criteria
2. psychological tests for selected candidates
3. second round of psychological tests and interviews
4. medical tests
5. job interview

The final list of applicants is submitted to the Director General of the European Space Agency for recruitment.

Where can I apply?

Registration takes place online and selected candidates are informed about the locations for the testing and interviews.

© Springer International Publishing AG 2017
E. Seedhouse, *Tim Peake and Britain's Road To Space*, Springer Praxis Books,
DOI 10.1007/978-3-319-57907-8

What are the required disciplines and qualifications to apply?
What should I study?

Typically, candidates should be knowledgeable in the scientific disciplines and should have demonstrated outstanding abilities in appropriate fields, preferably including operational skills.

Candidates can apply based on their academic or piloting credentials. Candidates submitting their applications based on their academic credentials should have a university degree (or equivalent), at the masters or preferably doctorate level, in the fields of the natural sciences (physics, biology, chemistry, and Earth sciences or related disciplines), medicine, engineering, information technology or mathematics. Flight and other operations experience are an asset.

Candidates submitting their applications on the basis of their qualifications as pilots will have logged a minimum of 1,000 hours in various high-performance aircraft. They should be test pilots and/or hold a university degree or equivalent (bachelors level or higher) in one of the aforementioned disciplines.

Successful candidates should have at least three years of experience in their respective professions and will have passed the necessary medical and psychological examinations, which will be administered during the astronaut selection process. It is a strong asset, but not mandatory, to have studied aeronautics and astronautics. Above all: no matter what you have studied, you should be good at it.

I don't speak English, should I apply?

This post requires a solid command of the English language, and it is an advantage to know another foreign language.

I don't speak Russian, should I apply?

Yes, speaking Russian is an asset but not a requirement. This is the second official language aboard the ISS, so you will learn it during astronaut training.

Which medical and psychological standards will be used to select the candidates?

An ESA Astronaut requires a multitude of skills, capabilities and characteristics. One important component of finding someone with the 'right stuff' is an assessment of how healthy each applicant is from a medical and psychological perspective.

In general, normal medical and psychological health standards are used. These standards are derived from evidence-based medicine, verified from clinical studies.

- An applicant should be able to pass a JAR-FCL 3, Class 2 medical examination or equivalent, conducted by an Aviation Medical Examiner certified by his/her national Aviation Medical Authority.

- The applicant must be free from any disease.
- The applicant must be free from any dependency on drugs, alcohol or tobacco.
- The applicant must have the normal range of motion and functionality in all joints.
- The applicant must have visual acuity in both eyes of 100 percent (20/20) either uncorrected, or corrected with lenses or contact lenses.
- The applicant must be free from any psychiatric disorders.
- The applicant must demonstrate cognitive, mental and personality capabilities to enable him/her to work efficiently in an intellectually and socially highly demanding environment

The rationale behind requiring an initial medical screening certificate will help ensure that astronaut applicants already possess a minimum level of health that is also required for anyone desiring a private pilot license.

The JAR-FCL 3 Class 2 (or equivalent such as those from the Federal Aviation Authority (FAA), European military or professional pilot medical certificate, or ESA Astronaut Applicant Medical Statement medical examination) is able to detect, efficiently and relatively inexpensively, many of the most common health-related factors that would prevent an applicant from becoming a private pilot, and hence also an astronaut.

This helps to streamline the process and ensure that applicants with the most likely chance for success are approved to continue to the next phase of the selection process. Obviously, there are more rigorous health demands to become an astronaut, hence the dedicated round of very intensive medical examinations that happen later in the selection process.

The JAR-FCL 3 Class 2 medical certificate is a European-wide accepted standard developed by the Joint Aviation Authority. It can only be issued by specifically certified physicians (i.e., aeromedical examiners). Recognizing that it may not be feasible for every interested applicant to obtain a JAR-FCL 3 Class 2 medical certificate, ESA will also accept an equivalent medical statement that has been authorized by a physician (for example a general medical practitioner).

The ESA Astronaut Medical Statement includes equivalent medical and health examinations to the JAR-FCL 3 Class 2, but can be authorized by a physician (e.g. general practitioner).

Remember, it is very important that you ONLY upload the Medical Certificate. Do NOT upload any medical data information associated with your Medical Certificate.

Do I need to be fit to become an astronaut? Which sport should I pursue?

It is important to be healthy, with an age-adequate fitness level. We are not looking for extreme fitness or top-level athletes – too many over developed muscles may be a disadvantage for astronauts in weightlessness.

There is no specific sport that can be recommended. Physical activities are in general beneficial to your health.

How can I prepare for the medical tests?

During the medical selection, applicants will undergo numerous tests across many health areas. Some tests are physically demanding, like bicycle or treadmill exercises. Some may also be invasive and others may be just questionnaires. There is nothing you can do generally to prepare for these medical examinations. If an examination requires specific preparations, such as fasting before giving a blood sample, applicants will receive instructions.

Do astronauts develop serious health problems during their stays in space?

No, there are no dangerous conditions that develop because of spaceflight. However, the space environment is hazardous and the astronauts' well-being depends on life-support systems. Weightlessness does have temporary potentially negative effects on human physiology, such as physical deconditioning and bone demineralization. The ESA Crew Medical Support Office and its staff are responsible for avoiding such hazards and preventing the space environment from affecting the physical and mental health of the astronauts. The environment and life-support systems are closely monitored, and there is a thorough preventive and countermeasure program.

Is it more difficult for a woman to become an astronaut?

No, from the physical point of view, it is not harder for a woman. The medical and psychological requirements for women and men are identical, apart from, of course, some gender-specific medical examinations.

Physical fitness and cardiovascular fitness are always evaluated on an individual basis and the fitness target values are adjusted to the physiological differences between men and women. A woman therefore does not have to meet the male norms, and vice-versa.

My vision is not perfect; can I still become an astronaut?

There is no clear yes/no answer because there is such a multitude of visual defects. However, vision problems account for most disqualifications. The main tests involve visual acuity, color perception and 3-D vision.

Wearing spectacles or contact lenses is not a reason for disqualification per se, but it has to be evaluated if, for example, a visual defect is known to progress rapidly. This could mean disqualification. Minor visual defects, even though requiring lenses, may be regarded as compatible with space duties.

Recently, a variety of surgical interventions to correct visual acuity has become more common. Some of these procedures will lead to disqualification, while others are acceptable. Every case will be judged individually.

In general, people applying to become European Astronauts (applicants) will need to submit many detailed pieces of information, participate in several rounds of psychological and medical screenings and participate in several interviews. What can you expect during the medical and psychological selection process when you apply to become an ESA Astronaut?

The psychological and medical selection is a rather lengthy process, covering a time span of about 10 months, separated into several phases. You will of course not be fully engaged during this entire time. Details about the time effort in each phase are provided in more detail below.

JAR-FCL 3, Class 2 medical examination certificate

One of the key first things you will need to do is to obtain a JAR-FCL 3, Class 2 medical examination certificate, or equivalent. This is a general physical examination performed by an authorized medical center under the authorization of your national aviation authority. Such a medical certificate is issued, for example, if you would like to become a glider pilot or a private pilot on a single-engine aircraft.

If an applicant does not meet the minimum medical requirements to become a private pilot, the more stringent medical requirements for an astronaut will definitely not be met. Therefore, this is the first medical screening step in the application process.

You can find more information about where to obtain this certificate from your national aviation authorities, on the internet and in local aviation clubs. Please be aware that you will have to cover the costs related to this examination.

You will need to upload a PDF copy of this JAR-FCL 3, Class 2 medical certificate, or equivalent, to the website. You should not upload any of your medical data, only the medical certificate is sufficient.

Medical questionnaires

Next, you'll be asked to provide answers to thorough medical questionnaires. You will be asked to provide answers to questions about your medical history, your current health status, your family medical history, your social habits and lifestyle, any history of disease in your family, etc. The answers you provide to this questionnaire will be strictly confidential and sent directly and only to the medical facility that will review this confidential information.

After a review of all online applications, there will be a selection of applicants to continue the process. At this point, if you are invited to the first round of psychological screening, you shall be referred to as a 'candidate'. Candidates will be invited in groups for a round of testing at a psychological center, called Psychological Stage 1.

At the same time you are invited as a candidate to Psychological Stage 1, your answers to the medical questionnaires will also be reviewed by the designated medical facility. Once at the psychology center, you will be assessed on your basic aptitudes and personality factors, using computer-based and/or paper tests in a group setting. You can assume a full-week effort on your side, including travel, for this stage of evaluations.

Candidates that pass this stage are invited to Psychological Stage 2.

Psychological Stage 2

The Psychological Stage 2 process focuses on evaluating you using behavior-oriented assessment in group exercises, role-playing, interviews and computer-based simulations. Again, you can expect roughly a one-week involvement on your side, including related travel to the testing site.

This stage verifies your personality and behavior in the context of the future profession, essentially verifying that you are qualified to perform the tasks and duties of an astronaut. Candidates that are positively reviewed in this stage may also begin cursory, non-invasive medical examinations via the Medical Selection process.

Medical examinations

After you uploaded your valid JAR-FCL 3, Class 2 medical examination certificate, or equivalent, you then provided answers to more specific medical questionnaires. This was reviewed while you were going through the Psychological Stages 1 and 2. If you have successfully passed these evaluations and the psychological tests, you will now go through a battery of medical examinations to assess your current health status and determine your relative risks for developing diseases in the future.

The purpose here is to screen for candidates that are free from disease, in excellent health and have the least likelihood for developing diseases in the future. In general, the examinations are designed from established standards in the following medical systems:

1. General Medicine
2. Ears, Nose, Throat
3. Ophthalmology
4. Pulmonology
5. Cardiovascular
6. Hematology
7. Abdomen and Digestive System
8. Endocrine and Metabolic
9. Genitourinary
10. Musculoskeletal and Orthopedics
11. Dermatology
12. Neurology
13. Psychiatry and Human Behavior
14. Obstetrics and Gynecology
15. Dental
16. Infectious Diseases
17. Anthropometry
18. Radiation Exposure
19. Nutrition
20. Physical Fitness
21. Special Tests as Indicated

This process will require a 10-day commitment (including travel) from you. There will be considerable care taken to make sure that you are not overly stressed or fatigued (e.g., jetlag after travel) during these medical examinations. For example, you will not be asked to perform a maximum effort stress test on a treadmill immediately after lunch. Additionally, test results may not be conclusive, so you may be asked to come back for follow-up tests, or to repeat a test to verify the results. Once all candidates have been medically evaluated, their confidential medical files will be sent to the ESA Medical Board.

ESA Medical Board

The ESA Medical Board will be comprised of numerous European medical experts representing the various medical disciplines you were tested in during the medical screening process. These experts will meet to review each candidate's medical test results. From this review, the ESA Medical Board will make recommendations for candidates to continue in the selection process. At this point in time, the psychological and medical selection process has come to an end.

Are there psychological and intellectual requirements?

General characteristics expected of applicants include, but are not limited to: good reasoning capability, the ability to work under stress, memory and concentration skills, aptitude for spatial orientation, psychomotor coordination and manual dexterity. An applicant's personality should be characterized by high motivation, flexibility, gregariousness, empathy with fellow workers, low level of aggression, and emotional stability.

What is the ideal age to apply?

The preferred age range is 27 to 37.

Is there a quota of astronauts for each ESA Member State?

No, there is no quota of astronauts. An effort is made to achieve, in the long run, a corps with astronauts from all ESA Member States.

For the selection process that included Tim Peake, ESA began its search for new astronauts on May 19, 2008, calling for applications from talented individuals who wished to join the European Astronaut Corps. Almost 10,000 individuals registered to begin the application process. At the close of the application phase, which lasted a month, 8,413 aspiring astronauts provided a medical certificate and finalized the online application form. This qualified them for the next step in the selection process.

Details

Most of the applications were received from France (22.1 percent) and Germany (21.4 percent), followed by Italy (11.0 percent), the United Kingdom (9.8 percent) and Spain (9.4 percent). Of the total number of applications, 16 percent were submitted by women.

ESA astronaut recruitment campaign 2008

Total number of applicants at the closure of the application period

Country	No. of Applicants	percentage of Total Applicants	As 2nd Citizenship	Men	Women
Austria	210	2.5	8	195	23
Belgium	253	3.0	8	224	37
Denmark	35	0.4	4	34	5
Finland	336	4.0	5	283	58
France	1860	22.1	58	1616	302
Germany	1798	21.4	35	1523	310
Greece	159	1.9	14	152	21
Ireland	128	1.5	11	110	29
Italy	927	11.0	39	815	151
Luxembourg	14	0.2	0	14	0
Norway	74	0.9	2	67	9
Other	72	0.9	309	301	80
Portugal	210	2.5	10	192	28
Spain	789	9.4	21	707	103
Sweden	172	2.0	9	156	25
Switzerland	351	4.2	26	325	52
The Netherlands	203	2.4	2	175	30
United Kingdom	822	9.8	42	697	167
Total	**8413**	**100**	**603**	**7586**	**1430**

ESA Astronaut Applicant Medical Examination List

The following is a list of required medical examinations that must be performed on a European Astronaut Applicant by a physician (e.g. general practitioner).

These medical examinations are equivalent to the JAR-FCL 3 Class 2 medical examinations.

ESA Medical Examinations and Parameters:

1. Interview/Questionnaire evaluation of family history, personal history and medical history

2. General physical examination, including

   a. all major organ systems, including skin
   b. mobility of extremities, joints and spine
   c. routine ears-nose-throat examination

      i. Hearing Analysis

   d. applicant must be able to understand correctly ordinary conversational speech at a distance of 2 meters from and with his back turned towards the examiner
   e. basic neurology assessment
   f. genito-urinary evaluation
   g. for females: gynecologic evaluation
   h. resting heart rate and blood-pressure

3. Standard 12-lead Resting ECG
4. Blood analysis for Hemoglobin (Hb), Lipids and Cholesterol
5. Urine stick analysis, including Glucose, Leucocytes, Erythrocytes, Protein
6. Ophthalmologic Analysis (can also be performed by ophthalmologist or vision care specialist, e.g. optometrist)

   a. vision in both eyes
   b. distant visual acuity with or without correction of 0.5 (6/12) or better in each eye separately and with both eyes 1.0 (6/6) or better

      i. Refractive error shall not exceed +5 to -8 diopters
      ii. Astigmatism shall not exceed 3 diopters

   c. normal color perception (Ishihara or Nagel's anomaloscope)

Appendix II

EVA tools

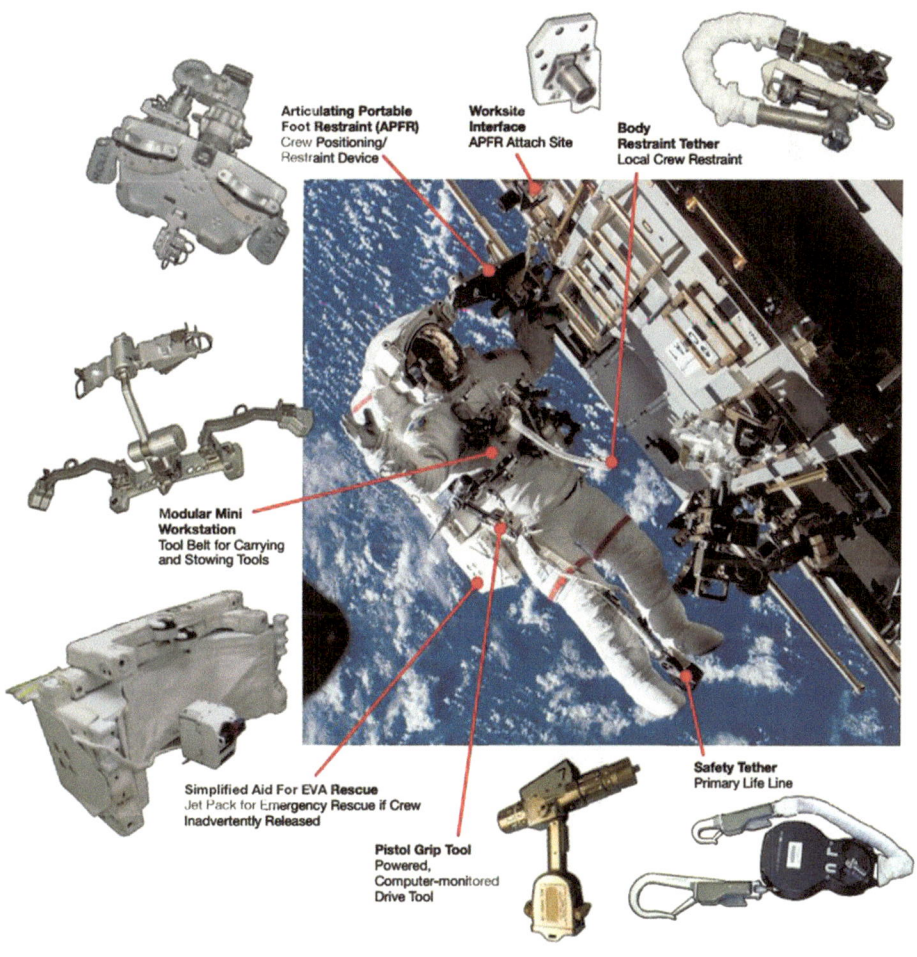

Articulating Portable
Foot Restraint (APFR)
Crew Positioning/
Restraint Device

Worksite
Interface
APFR Attach Site

Body
Restraint Tether
Local Crew Restraint

Modular Mini
Workstation
Tool Belt for Carrying
and Stowing Tools

Simplified Aid For EVA Rescue
Jet Pack for Emergency Rescue if Crew
Inadvertently Released

Pistol Grip Tool
Powered,
Computer-monitored
Drive Tool

Safety Tether
Primary Life Line

Credit: NASA

© Springer International Publishing AG 2017
E. Seedhouse, *Tim Peake and Britain's Road To Space*, Springer Praxis Books,
DOI 10.1007/978-3-319-57907-8

A selection of some of the tools astronauts use during EVAs

PGT (Pistol Grip Tool)
Basically, a fancy cordless drill, used for extraction and installation of bolts, and also used to torque bolts.

EMU Wrist Mirror
Extravehicular mobility unit wrist mirror, used to let the astronaut see things on their suit below their chest, since they can't bend their head forward and look.

RET (Retractable Equipment Tether)
One on PGT, one on bag. Whenever small items are to be transferred between crewman and crewman, or between crewman and bag, or between crewman and a structure, one of these tethers must be used to ensure the item doesn't float away.

Mini-Workstation Swing-Arm
Swings to stow itself up against the astronaut's side and provides additional bayonet receivers and clips for tools.

Six-foot Waist Tether
Used to anchor the astronaut at various sites on the structure, just to hold position. This is called *local tethering*, as opposed to more permanent attachment with the much larger 85-foot safety tether.

Bayonet Receivers
Fittings for attaching tools to the mini-workstation. Tools and equipment, like the EVA trash bag, have bayonet fittings that are mated to these receivers.

85-foot Safety Tether
When an astronaut egresses the airlock, this safety tether is immediately hooked to a designated position just outside the airlock. That way, the astronaut is always connected to the station no matter what.

End Effector — AKA "Grabber Daddy"
Another tethering device, used to grab onto a structure quickly and temporarily anchor an astronaut there. Not used as much as the waist tether or the BRT.

EVA Trash Bag
Small pouch with a double-flap opening. Gives the astronaut a place to stow bolts or other miscellaneous small items during the EVA so they don't float away.

D-Rings with D-Ring Extensions
The D-rings are used as attachment points for various things with clips, but the astronaut can't see them so they have *extenders* – the tethers coming down from them – which the astronaut can flip up and see.

BRT (Body Restraint Tether) clip
Secondary attachment point for the body restraint tether (BRT).

BRT (Body Restraint Tether) — stowed position
Flexible tool used to fix and hold an astronaut in position relative to a structure when he or she is going to perform a task. The BRT consists of an interlocking stack of spheres and can be rigidized to lock the astronaut into a position and/or orientation. It is also used to carry cargo such as tool bags, crew lock bags, or bigger bags that are clamped in the jaws and dragged behind the astronaut. The two brass-colored cylindrical knobs are used to adjust the BRT's joints once it is rigidized. The vertical knob rotates the elbow, and the horizontal knob rotates the wrist.

Appendix III

The Soyuz TMA Spacecraft

© Springer International Publishing AG 2017
E. Seedhouse, *Tim Peake and Britain's Road To Space*, Springer Praxis Books,
DOI 10.1007/978-3-319-57907-8

→ Soyuz - TMA

Russian manned spacecraft

Soyuz-TMA is a Russian manned spaceship capable of transporting up to three cosmonauts and limited cargo to and from the International Space Station.

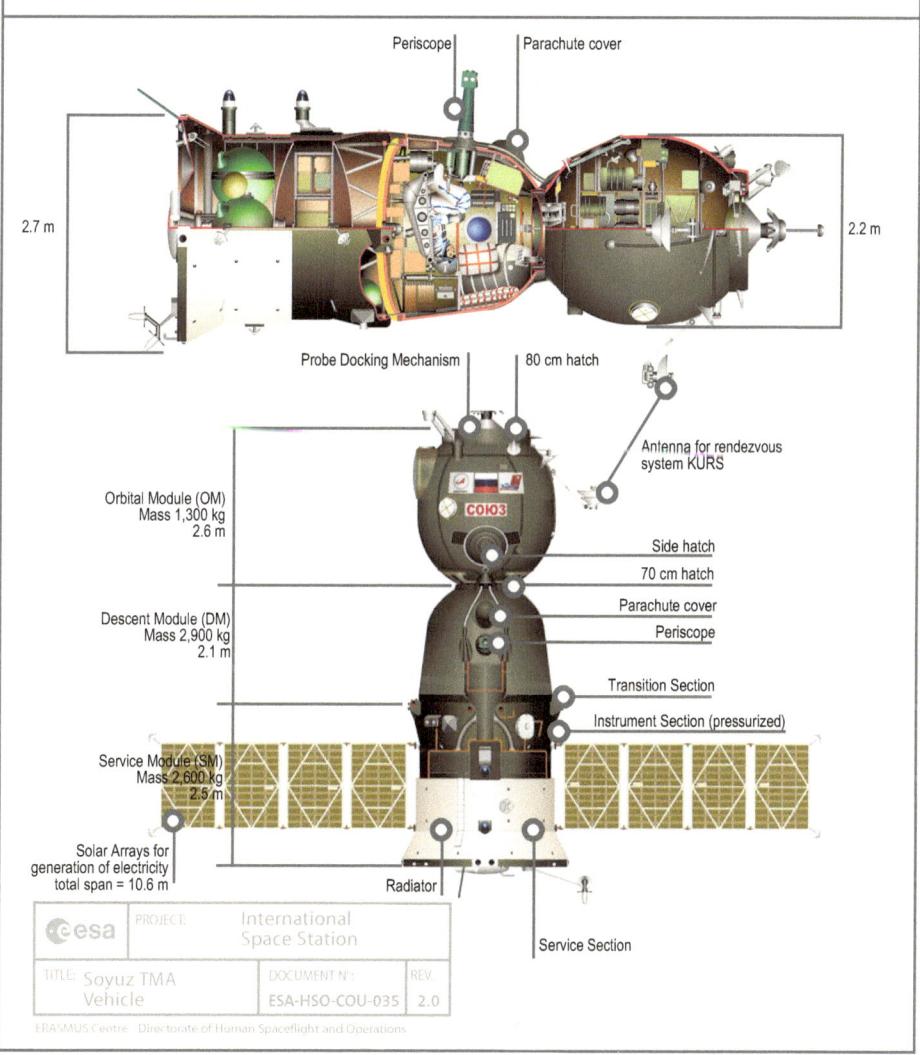

2.7 m

Periscope   Parachute cover

2.2 m

Probe Docking Mechanism   80 cm hatch

Antenna for rendezvous system KURS

Orbital Module (OM)
Mass 1,300 kg
2.6 m

СОЮЗ

Side hatch
70 cm hatch

Descent Module (DM)
Mass 2,900 kg
2.1 m

Parachute cover
Periscope

Transition Section
Instrument Section (pressurized)

Service Module (SM)
Mass 2,600 kg
2.5 m

Solar Arrays for
generation of electricity
total span = 10.6 m

Radiator

Service Section

esa	PROJECT:	International Space Station		
TITLE: Soyuz TMA Vehicle		DOCUMENT N°: ESA-HSO-COU-035	REV. 2.0	

ERASMUS Centre Directorate of Human Spaceflight and Operations

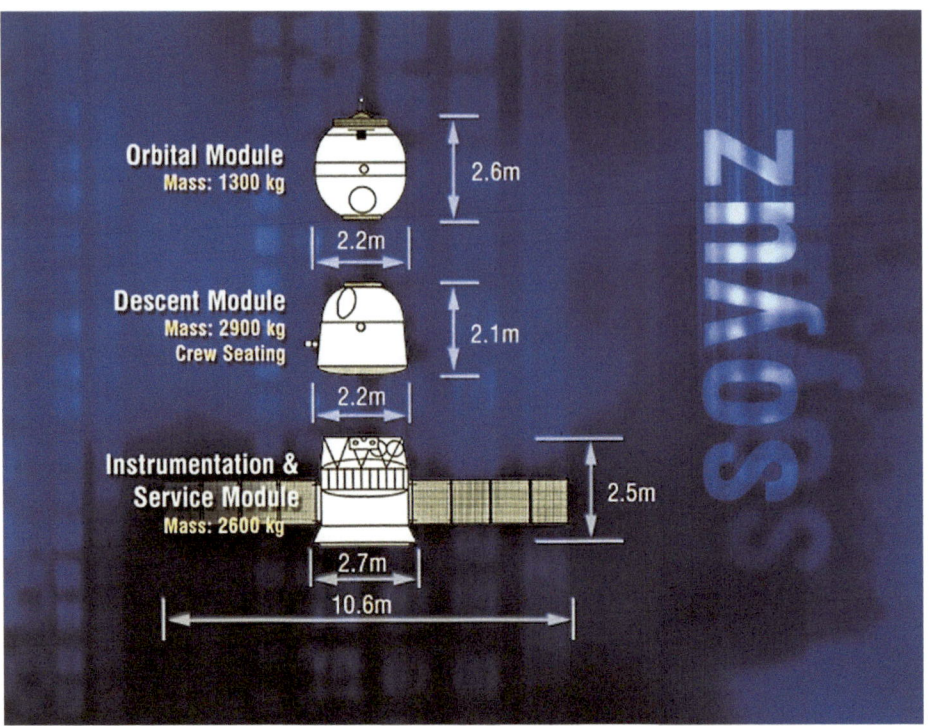

Orbital Module
Mass: 1300 kg

2.6m

2.2m

Descent Module
Mass: 2900 kg
Crew Seating

2.1m

2.2m

Instrumentation &
Service Module
Mass: 2600 kg

2.5m

2.7m

10.6m

Soyuz

Appendix IV

Expeditions 46 and 47

© Springer International Publishing AG 2017

E. Seedhouse, *Tim Peake and Britain's Road To Space*, Springer Praxis Books,

DOI 10.1007/978-3-319-57907-8

National Aeronautics and Space Administration

International Space Station
[M I S S I O N   S U M M A R Y]

EXPEDITION 46

Expedition 46 began December 11, 2015 and ends March 1, 2016. This expedition includes human research, biology and biotechnology, astrophysics research, physical science investigations and education activities. One spacewalk is tentatively planned during Expedition 46.

THE CREW:

Soyuz TMA-18M Launch: September 2, 2015 • Landing: March 1, 2016
(Note: Kelly and Kornienko launched on Soyuz TMA-16M on March 27, 2015)

Soyuz TMA-19M Launch: December 15, 2015 • Landing: June 5, 2016

Scott Kelly (NASA) – Flight Engineer

Born: Orange, New Jersey
Interests: racquetball, running, water sports and weight lifting
Spaceflights: STS-103, STS-118, Exps. 25 and 26
Bio: http://go.nasa.gov/SbcMZD
Twitter: @StationCDRKelly
Instagram: stationcdrkelly

Timothy Kopra (NASA) – Flight Engineer

Born: Austin, Texas
Interests: running, swimming, reading, home improvement projects, and spending time with family and friends
Spaceflights: STS-127, Expedition 20
Bio: http://go.nasa.gov/bgyJnW
Twitter: @astro_tim

Mikhail Kornienko (Roscosmos) – Flight Engineer
(Kor-knee-EHN-koh)

Born: Syzran, Russia
Interests: mountaineering
Spaceflights: Exps. 23 and 24
Bio: http://go.nasa.gov/TgOksk

Tim Peake (ESA) – Flight Engineer

Born: Chichester, England
Interests: skiing, scuba diving, cross-country running, climbing, and mountaineering
Spaceflights: Exps. 46/47 mark his first space station missions
Bio: http://go.nasa.gov/1MkB4Ja
Twitter: @astro_timpeake
Instagram: @astro_timpeake

Sergey Volkov (Roscosmos) – Flight Engineer
(SIR-gay VOLL-koff)

Born: Chuguyev, Kharkov Region, Ukraine
Interests: sports, tennis, windsurfing, reading and museums
Spaceflights: Exps. 17, 28 and 29
Bio: http://go.nasa.gov/1Os4JYn
Twitter: @Volkov_ISS

Yuri Malenchenko (Roscosmos) – Flight Engineer

Born: Svetlovodsk, Kirovograd Region, Ukraine
Interests: sports, games and music
Spaceflights: STS-106, Exps. 7, 16, 32 and 33
Bio: http://go.nasa.gov/195yzKl

THE SCIENCE:
What are some of the investigations the crew is operating?

During Expedition 46, crew members will install equipment and conduct experiments that help researchers test the use of inflatable space habitats in microgravity, analyze microbes living on the space station and study the protein crystal growth enablement of structure-based drug design (SBDD), an integral component in the drug discovery and development process. Investigations like these demonstrate how space station crews help advance NASA's journey to Mars while making discoveries that can benefit all of humanity.

International | **Mission**
Space Station | **Summary**

■ Bigelow Expandable Activity Module (BEAM)

Future space habitats for low-Earth orbit, the moon, Mars, or other destinations need to be lightweight and relatively simple to construct. The Bigelow Expandable Activity Module (BEAM) is an experimental expandable capsule that attaches to the International Space Station (ISS). After installation, BEAM inflates to roughly 13 feet long and 10.5 feet in diameter to provide a habitable volume where a crew member can enter.

Expandable habitats, occasionally described as inflatable habitats, greatly decrease the amount of transport volume for future space missions. These "expandables" weigh less and take up less room on a rocket while allowing additional space for living and working. They also provide protection from solar and cosmic radiation, space debris, and other contaminants. Crews traveling to the moon, Mars, asteroids, or other destinations could use them as habitable structures.

BEAM is scheduled to launch on SpaceX's Commercial Resupply Services Mission 8, also known as CRS-8. BEAM will be installed via the Canadarm2, which will remove BEAM from the capsule and connect it to the aft port of the International Space Station's Node 3. It will be inflated at a later date.

■ Microbial Tracking Payload Series/Microbial Observatory-1

Along with crew members and experimental payloads, the space station is home to a variety of microbes, which could potentially threaten crew health and jeopardize equipment. The Microbial Payload Tracking Series project uses microbial analysis techniques to establish a census of the microorganisms living on space station surfaces and in its atmosphere. Crew members will sample the United States modules three times during one year, which enables researchers to conduct long-term, multigenerational studies of microbial population dynamics. This analysis can help determine whether some microbes are more virulent in space, and which genetic changes might be involved in this response. This will provide a better understanding of microbe diversity onboard the station, as well as genetic strategies for identifying specific subsets. Results from this investigation can be used to evaluate cleaning strategies, and

to mitigate microbe-related risks to crew health and spacecraft system performance.

■ CASIS Protein Crystal Growth 4 (CASIS PCG 4)

CASIS PCG 4 comprises two investigations that both leverage the microgravity environment in the growth of protein crystals and focus on structure-based design (SBDD). SBDD is an integral component in the drug discovery and development process. Primarily, SBDD relies on the three-dimensional, structural information provided by protein crystallography to inform the design of more potent, effective and selective drugs.

One investigation will study the effect of microgravity on the co-crystallization of a membrane protein with a medically relevant compound. It has been established that growing protein crystals in microgravity can avoid some of the obstacles inherent to protein crystallization on Earth, such as sedimentation. As a result, scientists will attempt to grow co-crystals of a human membrane protein in the presence of a medically relevant compound in microgravity in order to determine its three-dimensional structure. This will enable scientists to chemically target and inhibit, with "designer" compounds, an important human biological pathway that has been shown to be responsible for several types of cancer.

The second investigation, A Co-Crystallization in Microgravity Approach to Structure-based Drug Design, seeks to determine whether crystals formed in microgravity represent an improvement over crystals formed by ground-based methods. Scientists expect the crystals formed in microgravity to diffract to a higher resolution than those developed on Earth, and thereby, provide greater molecular detail. This will permit more confident evaluations of ligand-binding (when a signal-triggering molecule binds to a site on a target protein). The resulting structures could be used to advance the medical-chemistry effort through improved/enhanced SBDD.

THE MISSION **PATCH:**

The 46 icon in the foreground of the patch represents the forty-sixth expedition on the International Space Station. Earth is depicted at the top with the flags of the countries of origin of the crew members: the United States, Russia and the United Kingdom. The Union flag of the UK is displayed in a position of prominence in recognition of the significance of the first British ESA (European Space Agency) astronaut to fly in space. The outer border is in the shape of a triangle with an unbroken border, symbolizing the infinite journey of discovery for past, present and future space explorers. The names of the six Expedition 46 astronauts and cosmonauts are shown in the border.

National Aeronautics and Space Administration

Lyndon B. Johnson Space Center
Houston, Texas 77058

www.nasa.gov

NP-2015-12-043-JSC

National Aeronautics and Space Administration

International Space Station

[M I S S I O N   S U M M A R Y]

EXPEDITION 47

Expedition 47 began March 1, 2016 and ends June 5, 2016. This expedition includes musculoskeletal research, chemistry research and a technology demonstration. No spacewalks are currently planned during Expedition 47.

THE CREW:

Soyuz TMA-19M Launch: December 15, 2015
(Note: Kelly and Kornienko launched on Soyuz TMA-16M on March 27, 2015) • Landing: June 1, 2016

Soyuz TMA-20M Launch: March 18, 2016 • Landing: September 7, 2016

Timothy Kopra (NASA) – Flight Engineer

Born: Austin, Texas
Interests: running, swimming, reading, home improvement projects, and spending time with family and friends
Spaceflights: STS-127, Expedition 20
Bio: http://go.nasa.gov/bgyJnW
Twitter: @astro_tim

Jeffrey Williams (NASA) – Flight Engineer

Born: Superior, Wisconsin
Interests: running, fishing, camping, skiing, scuba diving and woodworking
Spaceflights: STS-101, Exps. 13, 21 and 22
Bio: http://go.nasa.gov/20p7kDFI
Twitter: @Astro_Jeff
Instagram: @astro_jeffw

Timothy Peake (ESA) – Flight Engineer

Born: Chichester, England
Interests: skiing, scuba diving, cross-country running, climbing, and mountaineering
Spaceflights: Exps. 46/47 mark his first space station missions
Bio: http://go.nasa.gov/1MkB4Ja
Twitter: @astro_timpeake
Instagram: @astro_timpeake

Alexey Ovchinin (Roscomos) – Flight Engineer

Born: Rybinsk, Yaroslavl Region, Russia
Spaceflights: Exps. 47/48 mark his first space station missions
Bio: http://go.nasa.gov/20p7NFG

Yuri Malenchenko (Roscosmos) – Flight Engineer

Born: Svetlovodsk, Kirovograd Region, Ukraine
Interests: sports, games and music
Spaceflights: STS-106, Exps. 7, 16, 32 and 33
Bio: http://go.nasa.gov/195yzKl

Oleg Skripochka (Roscosmos) – Flight Engineer

Born: Nevinnomysk, Stavropol Region, Russia
Interests: parachute sport and bicycle tourism
Spaceflights: Exps. 25/26
Bio: http://go.nasa.gov/20p863z

THE SCIENCE:
What are some of the investigations the crew is operating?

During Expedition 47, researchers will investigate spaceflight's effect on the musculoskeletal system, the ability of tablets to dissolve in microgravity and how robotics can make exercise equipment smaller to minimize space dedicated to equipment and leave more room for crew during a long-duration mission. Investigations like these demonstrate how space station crews help advance NASA's journey to Mars while making discoveries that can benefit all of humanity.

International Space Station | Mission Summary

■ Rodent Research-3-Eli Lilly

Spaceflight causes a rapid loss of bone and muscle mass especially in the legs and spine, with symptoms similar to those experienced by people with muscle-wasting diseases or with limited mobility on Earth. Assessment of Myostatin Inhibition to Prevent Skeletal Muscle Atrophy and Weakness in Mice Exposed to Long-duration Spaceflight (Rodent Research-3-Eli Lilly), a U.S. National Laboratory investigation sponsored by the Center for the Advancement of Science in Space (CASIS), studies molecular and physical changes in the musculoskeletal system that happen in space. Results expand scientists' understanding of muscle atrophy and bone loss in space, while testing an antibody that has been known to prevent muscle wasting in mice on Earth.

In addition to the primary research focus on musculoskeletal systems, other organ systems are also studied for molecular and morphological changes as a function of duration of spaceflight exposure, further supporting the use of mice to model harmful effects of spaceflight in astronauts. On Earth, numerous diseases or physical impairments cause bone and muscle loss, including muscular dystrophy, cancer, spinal cord injury and the aging process. Patients on extended bed rest also experience similar physical changes. Results from this investigation could lead to new treatments for bone- and muscle-wasting diseases such as these.

■ Eli Lilly-Hard to Wet Surfaces

Another investigation hopes to determine how microgravity affects the ability of materials to dissolve. In chemistry, wetting refers to spreading of a liquid over a solid material's surface, and is a key aspect of the material's ability to dissolve. While tablets and pills that do not dissolve easily might impede a drug's release into the body, how a product's wettability affects its performance is not well understood. The Hard to Wet Surfaces (Eli Lilly-Hard to Wet Surfaces) investigation, also a U.S. National Laboratory study sponsored by CASIS, examines how certain materials used in the pharmaceutical industry dissolve in water while in microgravity.

On Earth, the density differences between a hard-to-wet solid/tablet and the solution can result in the solid/tablet floating on top of the solution, thereby exacerbating the dissolution problem. In microgravity, the solid/liquid density differences are negligible, and other factors controlling dissolution rate such as wettability dominate. Investigators hope to determine how mini-tablets behave differently in microgravity (float vs. sink, wet out faster or slower, etc.), and whether simple mixing will have less impact in microgravity (whether the tablet/capsule moves less). Results from this investigation could help improve the design of tablets that dissolve in the body to deliver drugs, thereby improving drug design for medicines used in space and on Earth.

■ Miniature Exercise Device (MED-2)

Exercise countermeasures are required by crew members during spaceflight to maintain health and to counter the debilitating effects of microgravity, including bone and muscle loss, cardiovascular alterations, and neurovestibular disturbances during long duration missions in microgravity. They are also especially critical for exploration missions, which require the crew to be at optimum physical performance in order to conduct potentially physically-demanding exploration tasks.

The current exercise equipment used on the International Space Station is large. Smaller exercise devices could make room for other critical spaceflight equipment while providing similar benefits. The Miniature Exercise Device (MED-2) technology demonstration exhibits key motion system technology required to reduce the volume and weight of countermeasure equipment that will be needed for long-term spaceflight. It demonstrates the use of robotic actuator technology to provide the motion and resistance needed to provide appropriate countermeasures for counteracting the effects of microgravity on the human body. This technology could lead to the next generation of exercise equipment that is lighter and smaller than existing systems and will be critical to longer duration spaceflight on journeys to Mars and beyond. Ground-based exercise equipment using the same robotic actuator technology could lead to improvements in rehabilitation and physical therapy, allowing physical therapists greater control over the prescriptions used in the exercises allowing for truly tailored rehabilitation programs.

THE MISSION **PATCH:**

The central depiction of the International Space Station (ISS) is in recognition of the international achievement of designing, building and maintaining a world-class space laboratory. The orientation of the ISS represents the view seen by the Soyuz crewmembers as they approach the station. The blackness of space in the background portrays the limitless area that humankind has yet to explore. The efforts of the Expedition 47 crew will contribute to the growing body of knowledge and expertise that will allow us to extend human exploration beyond low-Earth orbit. The three blue colors are from the flags of the Expedition 47 crew's home countries (United States, Russia and the United Kingdom), representing a fundamental commonality among each of the international partner countries whom the crewmembers serve.

National Aeronautics and Space Administration

Lyndon B. Johnson Space Center
Houston, Texas 77058

www.nasa.gov

NP-2016-02-004-JSC

Appendix V

PROGRESS

© Springer International Publishing AG 2017
E. Seedhouse, *Tim Peake and Britain's Road To Space*, Springer Praxis Books,
DOI 10.1007/978-3-319-57907-8

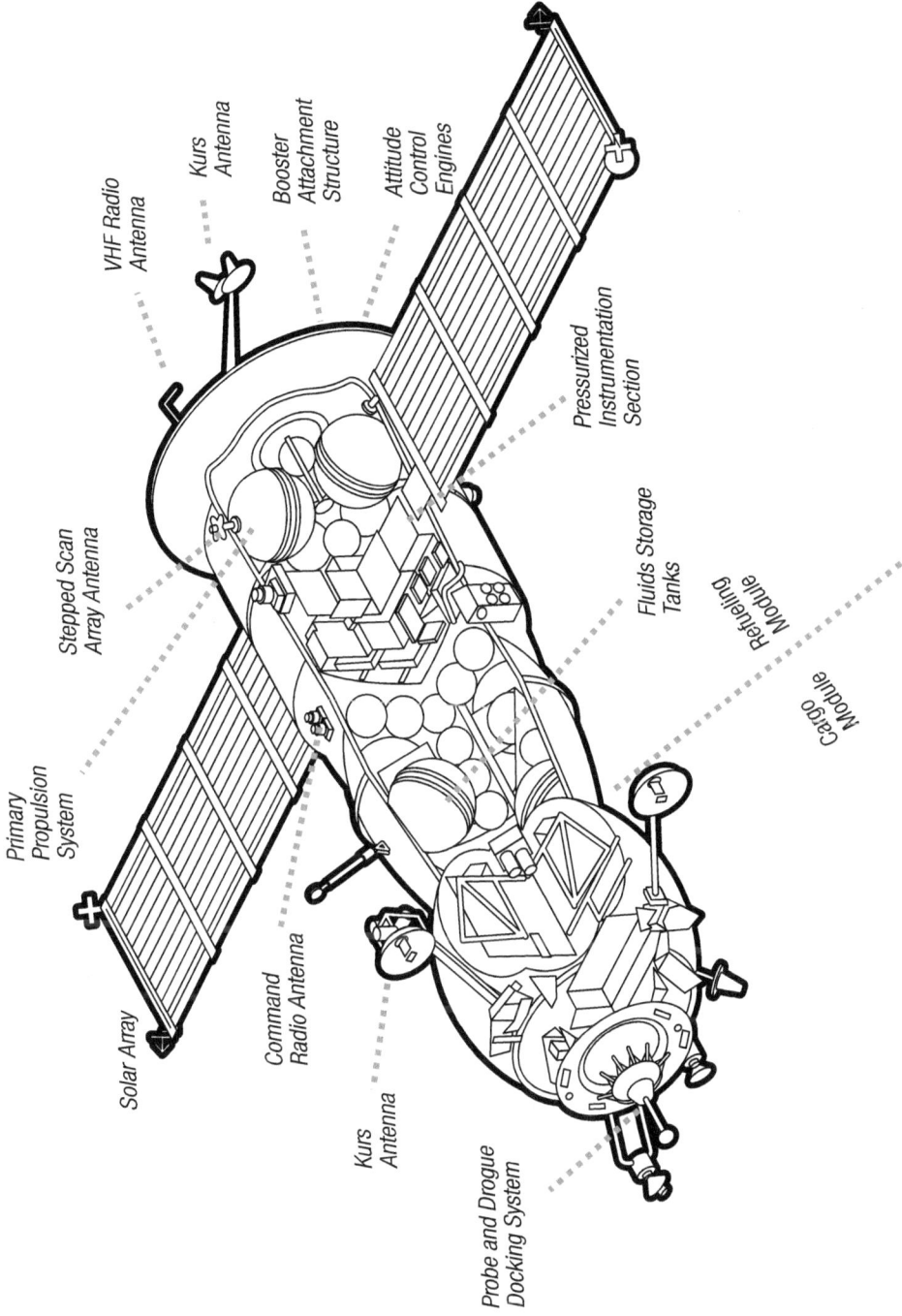

VHF Radio
Antenna

Kurs
Antenna

Booster
Attachment
Structure

Attitude
Control
Engines

Pressurized
Instrumentation
Section

Stepped Scan
Array Antenna

Fluids Storage
Tanks

Refueling
Module

Cargo
Module

Primary
Propulsion
System

Command
Radio Antenna

Solar Array

Kurs
Antenna

Probe and Drogue
Docking System

Appendix VI

Science on board the International Space Station

During their stay on board the ISS, the crews of Expeditions 46 and 47 carried out the following scientific experiments:

3D Printing in Zero-G (3D Printing in Zero-G Technology Demonstration)
3DA1 Camcorder (Panasonic 3D Camera)

ACE-H-2 (Advanced Colloids Experiment-Heated-2)
ACE-T-1 (Advanced Colloids Experiment-Temperature control-1)
AIRWAY MONITORING (AIRWAY MONITORING)
AMS-02 (Alpha Magnetic Spectrometer - 02)
APEX-04 (Epigenetic change in Arabidopsis thaliana in response to spaceflight - differential cytosine DNA methylation of plants on the ISS)
AQH Microscope Checkout (AQH Microscope Checkout)
ATOMIZATION (Detailed validation of the new atomization concept derived from drop tower experiments--Aimed at developing a turbulent atomization simulator)
Aniso Tubule (Roles of cortical microtubules and microtubule-associated proteins in gravity-induced growth modification of plant stems)
Area PADLES (Area Passive Dosimeter for Life-Science Experiments in Space)

BEAM (Bigelow Expandable Activity Module)
BP Reg (A Simple In-flight Method to Test the Risk of Fainting on Return to Earth After Long-Duration Space Flights)
Biochem Profile (Biochemical Profile)
Biological Rhythms 48hrs (The effect of long-term microgravity exposure on cardiac autonomic function by analyzing 48-hours electrocardiogram)
Bisphosphonates (Bisphosphonates as a Countermeasure to Space Flight Induced Bone Loss)
Body Measures (Quantification of In-Flight Physical Changes - Anthropometry and Neutral Body Posture)

© Springer International Publishing AG 2017
E. Seedhouse, *Tim Peake and Britain's Road To Space*, Springer Praxis Books,
DOI 10.1007/978-3-319-57907-8

CALET (CALorimetric Electron Telescope)
CARTILAGE (CARTILAGE)
CATS (Cloud-Aerosol Transport System)
Cardio Ox (Defining the Relation Between Biomarkers of Oxidative and Inflammatory Stress and Atherosclerosis Risk in Astronauts During and After Long-duration Spaceflight)
Circadian Rhythms (Circadian Rhythms)
Cognition (Individualized Real-Time Neurocognitive Assessment Toolkit for Space Flight Fatigue)

DOSIS-3D (Dose Distribution Inside the International Space Station - 3D)
Dose Tracker (Dose Tracker Application for Monitoring Medication Usage, Symptoms, and Adverse Effects During Missions)
Dynamic Surf (Experimental Assessment of Dynamic Surface Deformation Effects in Transition to Oscillatory Thermo capillary Flow in Liquid Bridge of High Prandtl Number Fluid)

ESA-Haptics-1 (ESA-Haptics-1)
Energy (Astronaut's Energy Requirements for Long-Term Space Flight)
ExHAM-Array Mark (On-orbit demonstration of target marker for space robotics)
ExHAM-CFRP Mirror (Space Environmental Testing of Lightweight and High-Precision Carbon Composite Mirrors)
ExHAM-MDM2 (Material Degradation Monitor on ExHAM)
ExHAM-PEEK (Space Environmental Testing of PEEK and PFA sample)
ExHAM-Solar Sail (Space Environment Exposure Tests of Functional Thin-Film Devices for Solar Sail)

FLEX-2J (Flame Extinguishment Experiment -2 JAXA)
Field Test (Recovery of Functional Sensorimotor Performance Following Long Duration Space Flight)
Fine Motor Skills (Effects of Long-Duration Microgravity on Fine Motor Skills: 1 year ISS Investigation)
Fluid Shifts (Fluid Shifts Before, During and After Prolonged Space Flight and Their Association with Intracranial Pressure and Visual Impairment)
Functional Task Test (Physiological Factors Contributing to Postflight Changes in Functional Performance)

Group Combustion (Elucidation of Flame Spread and Group Combustion Excitation Mechanism of Randomly-distributed Droplet Clouds)

HDEV (High Definition Earth Viewing)
HREP-HICO (HICO and RAIDS Experiment Payload - Hyperspectral Imager for the Coastal Ocean)
HREP-RAIDS (HICO and RAIDS Experiment Payload - Remote Atmospheric and Ionospheric Detection System (RAIDS))
Habitability (Habitability Assessment of International Space Station)
Hip QCT (Feasibility Study: QCT Modality for Risk Surveillance of Bone - Effects of In-flight Countermeasures on Sub-regions of the Hip Bone)

IMAX (IMAX Documentary film: A PERFECT PLANET)
IPVI (Non-invasive assessment of intracranial pressure for space flight and related visual impairment)
IPVI for 1YM (Non-invasive assessment of intracranial pressure for space flight and related visual impairment)
ISS External Leak Locator (ISS Robotic External Ammonia Leak Locator)
ISS Ham Radio (ARISS) (International Space Station Ham Radio (also known as Amateur Radio on the International Space Station (ARISS)))
ISS RapidScat (ISS-RapidScat)
Interfacial Energy 1 (Interfacial phenomena and thermophysical properties of high-temperature liquids-Fundamental research of steel processing using electrostatic levitation)
Intervertebral Disc Damage (Risk of Intervertebral Disc Damage after Prolonged Space Flight)

JAXA PCG (Japan Aerospace Exploration Agency Protein Crystal Growth)
JAXA PCG Demo (JAXA High Quality Protein Crystal Growth Demonstration Experiment)
Journals (Behavioral Issues Associated with isolation and Confinement: Review and Analysis of Astronaut Journals)

MAXI (Monitor of All-sky X-ray Image)
MCE (Multi-mission Consolidated Equipment)
MISSE-8 FSE (MISSE-8 FSE)
MUSCLE BIOPSY (MUSCLE BIOPSY)
MVIS Controller-1 (MVIS Controller-1)
Marangoni-UVP (Spatio-temporal Flow Structure in Marangoni Convection)
Medical Consumables Tracking (Medical Consumables Tracking)
Meteor (Meteor Composition Determination)
Micro-10 (Space Flight-Altered Motility Activation and Fertility-Dependent Responses in Sperm from Sea Urchin and Rodents)
Microbe-IV (Microbiological monitoring in the International Space Station-KIBO)
Microbial Observatory-1 (Microbial Tracking Payload Series)
Microbiome (Study of the Impact of Long-Term Space Travel on the Astronauts' Microbiome)
Multi-Omics (Multi-omics analysis of human microbial-metabolic cross-talk in the space ecosystem)

NanoRacks-Planet Labs-Dove (NanoRacks-Planet Labs-Dove)
NanoRacks-SyNRGE³ (NanoRacks-Symbiotic Nodulation in a Reduced Gravity Environment-Cubed)
NeuroMapping (Spaceflight Effects on Neurocognitive Performance: Extent, Longevity, and Neural Bases)

OPALS (Optical PAyload for Lasercomm Science)
Ocular Health (Prospective Observational Study of Ocular Health in ISS Crews)
OsteoOmics (Gravitational Regulation of Osteoblast Genomics and Metabolism)

PBRE (Packed Bed Reactor Experiment)

PK-4 (Plasma Krystall-4)

POP 3D (Portable Onboard Printer 3D)

PS-TEPC (Establishment of dosimetric technique in the International Space Station (ISS) with Position Sensitive Tissue Equivalent Proportional Chamber)

Plant RNA Regulation (Transcriptional and Post Transcriptional Regulation of Seedling Development in Microgravity)

Plant Rotation (Plant circumnutation and its dependence on the gravity response)

RJR (Augmented) Microbial Sampling (RJR (Augmented) Microbial Sampling)

RRM-Phase 2 (Robotic Refueling Mission Phase 2)

Radi-N2 (Radi-N2 Neutron Field Study)

Radiation Environment Monitor (Radiation Environment Monitor)

Reaction Self Test (Psychomotor Vigilance Self Test on the International Space Station)

Repository (National Aeronautics and Space Administration Biological Specimen Repository)

Robonaut (Robonaut)

SAGE III-ISS (Stratospheric Aerosol and Gas Experiment III/ISS)

SCAN Testbed (Space Communications and Navigation Testbed)

SEDA-AP (Space Environment Data Acquisition Equipment - Attached Payload)

SNFM (Serial Network Flow Monitor)

SPHERES-Zero-Robotics (Synchronized Position Hold, Engage, Reorient, Experimental Satellites-Zero-Robotics)

STMSat-1 (St. Thomas More School Cathedral Satellite-1)

Salivary Markers (The Effects of Long-Term Exposure to Microgravity on Salivary Markers of Innate Immunity)

Sally Ride EarthKAM (Sally Ride Earth Knowledge Acquired by Middle School Students)

Skin-B (Skin-B)

Sleep ISS-12 (Sleep-Wake Actigraphy and Light Exposure on ISS-12)

Solar-SOLACES (Sun Monitoring on the External Payload Facility of Columbus - SOLar Auto-Calibrating EUV/UV Spectrophotometers)

Solar-SOLSPEC (Sun Monitoring on the External Payload Facility of Columbus -Sun Monitoring on the External Payload Facility of Columbus -SOLar SPECtral Irradiance Measurements)

Space Headaches (Space Headaches)

Space Pup (Effect of space environment on mammalian reproduction)

Sprint (Integrated Resistance and Aerobic Training Study)

Stem Cells (Study on the Effect of Space Environment to Embryonic Stem Cells to Their Development)

Story Time From Space (Story Time From Space)

Synergy (The elucidation of the re-adaptation on the attitude control after return from long term space flight)

TBone (Assessment of the effect of space flight on bone quality using three-dimensional high resolution peripheral quantitative computed tomography (HR-pQCT))

Telomeres (Assessing Telomere Lengths and Telomerase Activity in Astronauts)

UBNT (Ultrasonic Background Noise Test)

V-C REFLEX (Plastic alteration of vestibulo-cardiovascular reflex and its countermeasure)
Vessel ID System (Vessel ID System)

Windows on Earth (Windows on Earth)

ZBOT (Zero Boil-Off Tank)

Eventually, the station command changed from U.S. astronaut Timothy Kopra to U.S. astronaut Jeffrey Williams. With undocking of Soyuz TMA-19M, carrying Yuri Malenchenko, Timothy Kopra and Timothy Peake, on June 18, 2016 at 05:52:33 UTC the Expedition 47 concluded and the new ISS Expedition 48 began.

During their stay on board the ISS, the crews of Expeditions 47 / 48 carried out the following scientific experiments (without Russian experiments):

3D Printing In Zero-G (3D Printing In Zero-G Technology Demonstration)

ACE-H-2 (Advanced Colloids Experiment-Heated-2)
ACE-T-1 (Advanced Colloids Experiment-Temperature control-1)
ACE-T-5-Bijels (Advanced Colloids Experiment-Temperature-5 Bijels)
AIRWAY MONITORING (AIRWAY MONITORING)
AMO-EXPRESS 2.0 (Autonomous Mission Operations EXPRESS 2.0 Project)
AMS-02 (Alpha Magnetic Spectrometer - 02)
APEX-04 (Epigenetic change in Arabidopsis thaliana in response to spaceflight - differential cytosine DNA methylation of plants on the ISS)
ARTE (Advanced Research Thermal Passive Exchange)
ATOMIZATION (Detailed validation of the new atomization concept derived from drop tower experiments-Aimed at developing a turbulent atomization simulator)
Area PADLES (Area Passive Dosimeter for Life-Science Experiments in Space)
At Home in Space (Culture, Values, and Environmental Adaptation in Space)
Auxin Transport (Studies on gravity-controlled growth and development in plants using true microgravity conditions)

BASS-II (Burning and Suppression of Solids - II)
BEAM (Bigelow Expandable Activity Module)
Biochem Profile (Biochemical Profile)
Biological Rhythms 48hrs (The effect of long-term microgravity exposure on cardiac autonomic function by analyzing 48-hours electrocardiogram)
Biomolecule Sequencer (Biomolecule Sequencer)
Bisphosphonates (Bisphosphonates as a Countermeasure to Space Flight Induced Bone Loss)
Body Measures (Quantification of In-Flight Physical Changes - Anthropometry and Neutral Body Posture)
Brain-DTI (Brain-DTI)

CALET (CALorimetric Electron Telescope)
CARTILAGE (CARTILAGE)

CASIS PCG 4-1 (Protein Crystallography to Enable Structure-Based Drug Design)
CASIS PCG 4-2 (The Effect of Microgravity on the Co-crystallization of a Membrane protein with a medically relevant compound.)
CASIS PCG 5 (Microgravity Growth of Crystalline Monoclonal Antibodies for Pharmaceutical Applications.)
CATS (Cloud-Aerosol Transport System)
CEO (Crew Earth Observations)
CFE-2 (Capillary Flow Experiment - 2)
CYTOSKELETON (CYTOSKELETON)
Cardio Ox (Defining the Relationship Between Biomarkers of Oxidative and Inflammatory Stress and the Risk for Atherosclerosis in Astronauts During and After Long-duration Spaceflight)
Cell Mechanosensing (Identification of gravity-transducers in skeletal muscle cells: Physiological relevance of tension fluctuations in plasma membrane)
Cell Science-01 (Cell Science-01)
Circadian Rhythms (Circadian Rhythms)
Cognition (Individualized Real-Time Neurocognitive Assessment Toolkit for Space Flight Fatigue)
Cool Flames Investigation (Cool Flames Investigation)

DECLIC DSI-R (DEvice for the study of Critical LIquids and Crystallization - Directional Solidification Insert-Reflight)
DECLIC HTI-R (DEvice for the study of Critical LIquids and Crystallization - High Temperature Insert-Reflight)
DOSIS-3D (Dose Distribution Inside the International Space Station - 3D)
Dose Tracker (Dose Tracker Application for Monitoring Medication Usage, Symptoms, and Adverse Effects During Missions)
Dynamic Surf (Experimental Assessment of Dynamic Surface Deformation Effects in Transition to Oscillatory Thermo capillary Flow in Liquid Bridge of High Prandtl Number Fluid)

ESA-EPO (European Space Agency-Education Payload Operation)
ESA-Haptics-1 (ESA-Haptics-1)
Eli Lilly-Hard to Wet Surfaces (Hard to Wet Surfaces)
Embryo Rad (Lifetime Heritable Effect of Space Radiation on Mouse embryos Preserved for a long-term in ISS)
Energy (Astronaut's Energy Requirements for Long-Term Space Flight)

FLEX-2 (Flame Extinguishment Experiment - 2)
Field Test (Recovery of Functional Sensorimotor Performance Following Long Duration Space Flight)
Fine Motor Skills (Effects of Long-Duration Microgravity on Fine Motor Skills: 1 year ISS Investigation)
Fluid Shifts (Fluid Shifts Before, During and After Prolonged Space Flight and Their Association with Intracranial Pressure and Visual Impairment)

Fruit Fly Lab -02 (FFL-02) (The effects of microgravity on cardiac function, structure and gene expression using the Drosophila model)

Gecko Gripper (Gecko Gripper)
Genes in Space-1 (Genes in Space-1)
Group Combustion (Elucidation of Flame Spread and Group Combustion Excitation Mechanism of Randomly-distributed Droplet Clouds)

HDEV (High Definition Earth Viewing)
HREP-RAIDS (HICO and RAIDS Experiment Payload - Remote Atmospheric and Ionospheric Detection System (RAIDS))
Habitability (Habitability Assessment of International Space Station)
Heart Cells (Effects of Microgravity on Stem Cell-Derived Heart Cells)

IPVI (Non-invasive assessment of intracranial pressure for space flight and related visual impairment)
ISS Ham Radio (ARISS) (International Space Station Ham Radio (also known as Amateur Radio on the International Space Station (ARISS)))
ISS RapidScat (ISS-RapidScat)
Immuno-2 (Immuno-2)
Interfacial Energy 1 (Interfacial phenomena and thermophysical properties of high-temperature liquids-Fundamental research of steel processing using electrostatic levitation)

JAXA ELF (Electrostatic Levitation Furnace (ELF))
JAXA PCG (JAXA PCG#11)
JAXA PCG Demo (JAXA High Quality Protein Crystal Growth Demonstration Experiment)

LDST (Long Duration Sorbent Testbed)
LMM Biophysics 1 (The Effect of Macromolecular Transport of Microgravity Protein Crystallization)
LMM Biophysics 3 (Growth Rate Dispersion as a Predictive Indicator for Biological Crystal Samples Where Quality Can be Improved with Microgravity Growth)
LONESTAR (Low Earth Orbiting Navigation Experiment for Spacecraft Testing Autonomous Rendezvous and Docking)

MAGVECTOR (MAGVECTOR)
MAXI (Monitor of All-sky X-ray Image)
MED-2 (Miniature Exercise Device)
METERON (METERON Quick Start a / DTN)
MISSE-8 FSE (MISSE-8 FSE)
MUSCLE BIOPSY (MUSCLE BIOPSY)
MVIS Controller-1 (MVIS Controller-1)
Marangoni-UVP (Spatio-temporal Flow Structure in Marangoni Convection)
Maritime Awareness (Global AIS on Space Station (GLASS))
Marrow (The MARROW study (Bone Marrow Adipose Reaction: Red Or White?))
Meteor (Meteor Composition Determination)

Micro-10 (Influence of microgravity on the production of Aspergillus secondary metabolites (IMPAS) – a novel drug discovery approach with potential benefits to astronauts' health)
Micro-9 (Yeast colony survival in microgravity depends on ammonia mediated metabolic adaptation and cell differentiation)
Microbe-IV (Microbiological monitoring in the International Space Station-KIBO)
Microbial Observatory-1 (Microbial Tracking Payload Series)
Microbiome (Study of the Impact of Long-Term Space Travel on the Astronauts' Microbiome)
Microchannel Diffusion (Microchannel Diffusion)
Mouse Epigenetics (Transcriptome analysis and germ-cell development analysis of mice in space)
Multi-Omics (Multi-omics analysis of human microbial-metabolic cross-talk in the space ecosystem)
Myco (for 1YM) (Mycological Evaluation of Crew Exposure to ISS Ambient Air 1 Year Mission)

NanoRacks-AGAR (NanoRacks-Algal Growth and Remediation)
NanoRacks-Gumstix (NanoRacks-Evaluation of Gumstix Performance in Low-Earth Orbit)
NanoRacks-JAMSS-2, Lagrange-1 (NanoRacks-JAMSS-2, Lagrange-1)
NanoRacks-LEMUR-2 (NanoRacks-LEMUR-2)
NanoRacks-Mission Discovery 2 (NanoRacks–Mission Discovery Biomedical Experiments 2)
NanoRacks-NCESSE-Odyssey (NanoRacks-National Center for Earth and Space Science-Odyssey (SSEP Mission 7))
NanoRacks-NovaWurks-SIMPL-Microsat (NanoRacks Kaber Mission 1-NovaWurks-Satlet Initial Mission Proofs and Lessons)
NanoRacks-Slime Mold (NanoRacks-Slime Mold Organization)
NanoRacks-SyNRGE[3] (NanoRacks-Symbiotic Nodulation in a Reduced Gravity Environment-Cubed)
NeuroMapping (Spaceflight Effects on Neurocognitive Performance: Extent, Longevity, and Neural Bases)

OASIS (Observation Analysis of Smectic Islands in Space)
OPALS (Optical PAyload for Lasercomm Science)
Ocular Health (Prospective Observational Study of Ocular Health in ISS Crews)

PBRE (Packed Bed Reactor Experiment)
PK-4 (Plasma Krystall-4)
Personal CO2 Monitor (Personal CO2 Monitor)
Phase Change HX (Phase Change Heat Exchanger Project)
Plant Gravity Sensing (Utilization of the micro gravity condition to examine the cellular process of formation of the gravity sensor and the molecular mechanism of gravity sensing)
Plant RNA Regulation (Transcriptional and Post Transcriptional Regulation of Seedling Development in Microgravity)

REBR-W (Reentry Breakup Recorder with Wireless Sensors)
RFID Logistics Awareness (RFID-Enabled Autonomous Logistics Management (REALM))
RJR (Augmented) Microbial Sampling (RJR (Augmented) Microbial Sampling)
ROSA (Roll-Out Solar Array)
RRM-Phase 2 (Robotic Refueling Mission Phase 2)
RTcMISS (Radiation Tolerant Computer Mission on the ISS)
Radi-N2 (Radi-N2 Neutron Field Study)
Radiation Environment Monitor (Radiation Environment Monitor)
Repository (National Aeronautics and Space Administration Biological Specimen Repository)
Robonaut (Robonaut)
Rodent Research-3-Eli Lilly (Assessment of myostatin inhibition to prevent skeletal muscle atrophy and weakness in mice exposed to long-duration spaceflight)
Rodent Research-4 (CASIS) (Tissue Regeneration-Bone Defect)

SAGE III-ISS (Stratospheric Aerosol and Gas Experiment III-ISS)
SCAN Testbed (Space Communications and Navigation Testbed)
SEDA-AP (Space Environment Data Acquisition Equipment - Attached Payload)
SNFM (Serial Network Flow Monitor)
SODI-DCMIX (SODI-DCMIX)
SPHERES Halo (Synchronized Position, Hold, Engage, Reorient, Experimental Satellites - Halo)
SPHERES Tether Demo (SPHERES Tether Demo)
SPHERES-Slosh (SPHERES-Slosh)
SPHERES-UDP (Synchronized Position, Hold, Engage, Reorient, Experimental Satellites-Universal Docking Port)
SPHERES-Zero-Robotics (Synchronized Position Hold, Engage, Reorient, Experimental Satellites-Zero-Robotics)
SPHEROIDS (SPHEROIDS)
STP-H5 FPS (STP-H5-Fabry Perot Spectrometer for Methane)
STP-H5 ICE (STP-H5-Innovative Coatings Experiment)
STP-H5 LITES (STP-H5-Limb-Imaging Ionospheric and Thermospheric Extreme-Ultraviolet Spectrographs)
STP-H5 SHM (STP-H5-Structural Health Monitoring)
STP-H5 Space Cube - Mini (STP-H5-SpaceCube - Mini)
Saffire-I (Spacecraft Fire Experiment-I)
Saffire-II (Spacecraft Fire Experiment-II)
Salivary Markers (The Effects of Long-Term Exposure to Microgravity on Salivary Markers of Innate Immunity)
Sally Ride EarthKAM (Earth Knowledge Acquired by Middle School Students)
Skin-B (Skin-B)
Solar-SOLACES (Sun Monitoring on the External Payload Facility of Columbus - SOLar Auto-Calibrating EUV/UV Spectrophotometers)

Solar-SOLSPEC (Sun Monitoring on the External Payload Facility of Columbus-Sun Monitoring on the External Payload Facility of Columbus-SOLar SPECtral Irradiance Measurements)
Space Headaches (Space Headaches)
Space Pup (Effect of space environment on mammalian reproduction)
Sprint (Integrated Resistance and Aerobic Training Study)
Stem Cells (Study on the Effect of Space Environment to Embryonic Stem Cells to Their Development)
Story Time From Space (Story Time From Space)
Straight Ahead in Microgravity (Straight Ahead)
Strata-1 (Strata-1)
Synergy (The elucidation of the re-adaptation on the attitude control after return from long term space flight)
Synthetic Muscle (Synthetic Muscle: Resistance to Radiation; Ras Labs-CASIS-ISS Project for Synthetic Muscle: Resistance to Radiation)

TBone (Assessment of the effect of space flight on bone quality using three-dimensional high resolution peripheral quantitative computed tomography (HR-pQCT))
Telescience Resource Kit (Flight Demonstration of Telescience Resource Kit)
Telomeres (Assessing Telomere Lengths and Telomerase Activity in Astronauts)
Try Zero-G for Asia (Try Zero-G for Asia)

UBNT (Ultrasonic Background Noise Test)
UD Space Suit Layup (Improved EVA Suit MMOD Protection using STF-Armor™ and self-healing polymers)
Universal Battery Charger (Universal Battery Charger)

Vascular Echo (Cardiac and Vessel Structure and Function with Long-Duration Space Flight and Recovery)
Veg-03 (Veg-03)
Vessel ID System (Vessel ID System)

WISENET (WISENET)
Water Monitoring Suite (Water Monitoring Suite)
Windows on Earth (Windows on Earth)

ZBOT (Zero Boil-Off Tank)

Rocket Science Experiment

Rocket Science is giving 10,000 UK schools the opportunity to engage their pupils in a UK-wide live science experiment to contribute to our knowledge of growing plants in space. After participating in a classroom experiment in May and June 2016, pupils will be asked to enter their results in a bespoke microsite so that results from schools across the nation can be collated and analysed by professional biostatisticians.

Two kilograms of rocket seeds (*Eruca sativa*) were launched on Soyuz 44S on 2 September 2015 with European Space Agency (ESA) astronaut Andreas Mogensen and his

Experiment Overview

crew, arriving on the International Space Station (ISS) two days later. British ESA astronaut Tim Peake will take charge of the seeds while on the ISS for his Principia mission starting in December. After being held for about six months in microgravity, the seeds will be returned to Earth with astronaut Scott Kelly, currently planned for March 2016.

Once the seeds have returned they will then be distributed to schools signed up to the project. Each participating school will receive 100 seeds that have been on the ISS and 100 seeds that have remained on Earth. The seed packets will be colour coded, however schools will not be told which packet contains which seeds until national results have been published. Online resources to expand student learning will be available on the website of the UK Space Education Office (esero.org.uk) before, during and after the Rocket Science experiment.

Full instructions will be provided in a printed Rocket Science Teachers' Pack, which will be suitable for all ages and distributed to schools by the Royal Horticultural Society (RHS) before April 2016, ready to begin the experiment in May. The pack will contain the two packets of rocket seeds, an A1-sized wall chart to enable pupils to monitor seedling growth, fun stickers and a booklet with simple step by step instructions for the experiment and room to record data.

Equipment

Teachers will be asked to source the following equipment before the arrival of the seeds in April 2016:

- 1 x 25 litre bag of compost (1x 20 litres fills 8 trays).
- 8 x seed module trays with a space for each of the 200 seeds e.g. a P40 tray that has 40 individual square cells for each seed. Please see the image below.
- Clean and empty clear milk cartons for labels. Each seed needs to be labelled, and for this you will require 8 x 2 pints, 4 x 1 litre or 2 x 2 litre cartons
- Classroom space to house the experiment (a single location, preferably with a good source of light but away from direct sunlight, radiators etc.)

A group of pupils will need to be identified to perform the experiment. This could be a class, STEM club, gardening club or other group. It is envisaged that this group will also present its findings to the wider school, for example through an assembly.

Learning objectives

1. Students will work scientifically to compare the growth of the two types of seeds through research and observation, suggest which is which and suggest possible explanations for their results. Some will also practise effective science communication skills when presenting to the wider school.
2. Students will understand the components of good plant growth i.e. soil, water and light and be introduced to new components such as the presence of gravity.
3. Students may also be inspired to find out more about careers in STEM subjects including horticulture, plant science and the space industry as a whole.

Methods

Each seedling will have a letter and number code. Pupils will take 9 measurements over 35 days collecting data as instructed on germination, growth, leaf count and plant height at frequent intervals. Randomisation is a key element of the experiment process to ensure that the data is unbiased. A simple explanation of how to achieve this will be provided in the Teachers' Pack. The data can quickly be recorded on the Rocket Science wall chart in the classroom and the experiment booklet, and entered at a later date on the data collection website following instructions that will be provided.

Seedling Care

The rocket seeds will be sown on the first available Monday of the summer term following receipt (Day 1). The experiment is organised into 7 weeks of work (totalling 35 days) and no weekend work is necessary. Your seedlings will need to be checked daily for watering, so appoint a pupil each day to do this. For the weekend ensure the seedlings are watered before you go home and then check them first thing on Monday morning. Pupils will gain key horticultural skills such as seed sowing and watering as a result of their participation in Rocket Science.

Results

After all the data has been collected, the results will be analysed by professional statisticians. Leading scientists from the RHS and European Space Agency will interpret the results and draw possible conclusions. An online report will also be made available on the RHS Campaign for School Gardening website from September 2015.

Index

© Springer International Publishing AG 2017
E. Seedhouse, *Tim Peake and Britain's Road To Space*, Springer Praxis Books,
DOI 10.1007/978-3-319-57907-8